Sustainable Visions: Off-Grid Projects for the Modern Homesteader

Elevate Your Lifestyle with Green Innovations

Lily Martin

Table of Contents

INTRODUCTION

"Sustainable Visions: Off-Grid Projects for the Modern Homesteader - Elevate Your Lifestyle with Green Innovations" is a groundbreaking e-book that provides a comprehensive guide for individuals seeking to embrace a sustainable and off-grid lifestyle. Authored by experts in the field of eco-friendly living, this e-book is a treasure trove of insights, practical tips, and innovative projects designed to empower modern homesteaders to live harmoniously with nature.

In a world where environmental concerns and the quest for self-sufficiency are becoming increasingly prevalent, "Sustainable Visions" emerges as a beacon of inspiration and knowledge. The e-book begins by exploring the fundamental principles of sustainability, laying the groundwork for readers to understand the importance of reducing their ecological footprint and adopting eco-conscious practices. From there, it seamlessly transitions into a wealth of off-grid projects that can be easily implemented by anyone aspiring to lead a more sustainable lifestyle.

Readers will find a diverse range of projects within the pages of this e-book, encompassing energy generation, water conservation, waste reduction, and sustainable agriculture. Whether harnessing solar power, creating DIY rainwater harvesting systems, implementing permaculture techniques, or constructing eco-friendly dwellings, each project is meticulously explained with step-by-step instructions and accompanied by vivid illustrations.

What sets "Sustainable Visions" apart is its holistic approach, addressing the technical aspects of off-grid living and the broader philosophy behind it. The authors delve into the mindset shift required to embrace a sustainable lifestyle fully, emphasizing the interconnectedness of personal choices with the planet's well-being.

As readers embark on this enlightening journey through the e-book, they will be equipped with the knowledge and motivation to transform their homesteads into thriving, eco-friendly havens. "Sustainable Visions" is more than just a guide; it is a manifesto for a greener, more sustainable future, inviting readers to join the ranks of the modern homesteading movement.

CHAPTER I

Setting the Foundation

Assessing your land for off-grid living

Embarking on the journey of off-grid living is a transformative endeavor that demands careful consideration and planning, starting with a thorough assessment of the land. The success of any off-grid project hinges on the understanding and utilization of the natural resources available. This section explores the critical elements of determining if a property is suitable for off-grid living, including topographical analysis, climatic concerns, water supplies, and the possibilities for renewable energy.

An in-depth study of the terrain is crucial, first and foremost. The land's topography is a significant factor in deciding if off-grid living is feasible. Construction and farming may be more difficult on steep slopes, but flat areas provide greater flexibility. Another important consideration is accessibility; although remote areas can offer seclusion, they can also pose logistical difficulties. Thriving agriculture requires an analysis of the soil composition since different crops do better in various soil types. A soil test can provide information on the amount of nutrients present, the soil's capacity for drainage, and any obstacles that can arise when cultivating crops.

The climate is an essential factor to consider while evaluating land for off-grid life. Temperature ranges, precipitation quantities, and sunlight exposure differ throughout areas. It is necessary to comprehend the climate trends of a particular place when designing energy systems, managing water resources, and choosing appropriate crops. Living rooms may need more insulation in colder climates, and efficient cooling techniques may be required in hotter climates. Rainfall patterns influence water availability, and the sustainability of solar energy systems is impacted by sunlight exposure. Off-grid enthusiasts can adapt their techniques to the specific features of the area by doing thorough climate analysis, which ensures resistance against climatic problems.

Since water sources are essential to off-grid life, determining water availability and quality is crucial to assessing a piece of property. Access to a dependable and sustainable water source is a must for off-grid life. Familiar sources such as rivers, wells, springs, and rainwater collection systems should be assessed for their dependability and closeness to the residential area. Testing the water's quality is necessary to ensure it's safe to drink and use for other household needs. Furthermore, since municipal laws governing water rights and usage differ by area, it is crucial to comprehend them.

Off-grid living's potential for renewable energy is essential, and determining if a piece of land is suitable for capturing solar, wind, or hydroelectric power is crucial. Because solar panels need lots of sunshine, it's essential to determine how many hours, on average, there are of sun each day of the year. Wind turbines should be installed with consideration for places with moderate to constant wind speeds. Hydroelectric systems could be a good choice for people living close to water bodies. Off- grid enthusiasts can optimize their energy systems and guarantee a dependable and sustainable power source by conducting a renewable energy assessment.

When evaluating land for off-grid living, it's equally critical to consider wildlife and biodiversity. In addition to improving the surroundings' aesthetic appeal, a rich and diversified ecology is a sign of a healthy environment. Comprehending the indigenous flora and wildlife is vital for sustainable cohabitation. It facilitates the identification of any obstacles, such as wildlife, that could endanger crops, and it encourages the use of environmentally friendly methods that protect and improve the region's biodiversity.

Legal and zoning issues are essential when assessing land for off-grid living. Zoning laws, building codes, or local regulations may restrict specific projects or activities. Preliminary research into these areas helps avoid legal entanglements and guarantees that off-grid operations adhere to applicable laws. Furthermore, it is essential to comprehend property lines and possible easements to prevent disagreements and confrontations with nearby landowners.

To sum up, determining whether a property is suitable for off-grid living requires careful evaluation of several elements, including the terrain, climate, water availability, possibilities for renewable energy production, wildlife and biodiversity, and legal issues. An all- encompassing method of evaluating land guarantees that those interested in living off the grid are knowledgeable and prepared to make choices supporting their environmental objectives. Giving these factors careful thought makes off-grid living more feasible and helps us live in harmony with the earth. A successful and satisfying off-grid lifestyle is contingent upon a thorough assessment of the land, which is becoming increasingly important as people look for alternatives to traditional living.

Choosing the right location

Selecting the ideal location is a pivotal decision for individuals contemplating a shift towards off-grid living. The correct location is the foundation for the off-grid lifestyle, influencing everything from sustainability and self-sufficiency to the overall quality of life. This section explores the various issues people should consider when deciding where to live off the grid. It looks at things like climate, resource accessibility, community dynamics, legal considerations, and the psychological effects of the site.

The climate is a significant factor in deciding whether off-grid living is feasible and comfortable. The weather, temperatures, and seasonal fluctuations that differ between areas significantly impact day-to-day living. It is necessary to know the climate of a prospective site to plan energy systems, agricultural practices, and even architectural designs. Places with severe winters need more insulation and heating options, whereas hot summers might need efficient cooling techniques. The viability of solar panels and rainwater collection systems— two crucial elements of sustainable off-grid living—is significantly impacted by the amount of precipitation and sunshine available. Thus, it is essential to carefully consider a potential place's climate to ensure that it meets the needs and tastes of people who want to live off the grid.

Selecting a site for off-grid living requires careful consideration of accessibility to necessary resources. Sufficient and sustainable water sources are essential for life and various everyday tasks. Prospective off-grinders should evaluate the quality and quantity of water sources that they may obtain, such as rainwater harvesting systems, wells, springs, and rivers. Being close to these sources is crucial since accessibility and transportation affect how feasible it is to use these resources. Similarly, determining whether the property has fertile soil supporting agriculture is vital for anyone hoping to produce enough food. The practicality of wood for cooking and heating may be impacted by a location's proximity to wood or other biomass sources. An in-depth analysis of the resources at hand guarantees that off-grid life is feasible and sustainable in the selected area.

Social and community dynamics are essential factors that can significantly impact the off-grid experience. Although people are drawn to isolated areas by the desire for independence and seclusion, it is necessary to recognize the value of neighboring communities. Building relationships with neighbors who share your values promotes community support and establishes a network for exchanging information, resources, and help in trying times. Comprehending the customs and interpersonal relationships of the area enables off-grid enthusiasts to blend in with their surroundings with ease, which enhances the overall satisfaction and fulfillment of living off the grid.

When choosing an off-grid site, regulations and legal issues should be considered. Zoning laws, building rules, and other regulations differ throughout locations and might affect the viability of particular projects or activities. Off-grid enthusiasts should familiarize themselves with these legal frameworks and conduct careful studies to ensure their ideas comply with local laws. This includes getting building permits, handling trash disposal laws, and abiding with land-use constraints. Ignoring these factors might cause legal issues and make implementing off-grid initiatives more difficult. Therefore, a thorough awareness of the legal environment is essential for people looking to live off the grid.

A less obvious but no less significant component is the chosen place's psychological effect on people interested in off-grid life. A place's natural surroundings, architecture, and general atmosphere all impact its people's mental and emotional health. A higher quality of life can be attributed to elements that favor mental health, such as beautiful scenery, being close to nature, and being free from urban noise. On the other hand, solitude and harsh surroundings can be detrimental to mental health. Thus, people must evaluate how well the place suits their psychological requirements and preferences.

In addition, the decision-making process needs to consider healthcare and education. People who have children should assess the caliber and accessibility of local educational institutions. Similarly, providing off-grid dwellers access to medical facilities and emergency services is essential to their well-being. It's a delicate endeavor that necessitates careful analysis of possible places to balance the need for solitude and pragmatic concerns like healthcare and education.

In summary, selecting the ideal site for off-grid living is a complex process that necessitates thoroughly examining the local climate, the accessibility of resources, community dynamics, legal issues, and the psychological effects of the chosen environment. The choice requires striking a careful balance between sustainability, the need for self-sufficiency, and the pragmatics of everyday life. Living off the grid successfully involves more than just choosing a place to live; it also involves designing a happy, satisfying lifestyle that fits one's personal goals and values. By carefully weighing these many elements, people can make decisions that set the stage for a fulfilling and sustainable off-grid lifestyle.

Legal considerations and permits

Embarking on the journey of off-grid living is a transformative decision that involves more than just embracing a self-sufficient lifestyle—it necessitates a deep understanding of the legal landscape and compliance with regulations governing land use, construction, and environmental practices. This section addresses the complex issues people must handle to ensure a legal and successful transition to an off-grid lifestyle, emphasizing the vital significance of legal considerations and permits in the off-grid living space.

Zoning and land-use regulations are two of the most critical legal factors in off-grid living. Zoning laws vary by area and specify the uses of land that can be made of it, from commercial and industrial to residential and agricultural. Off-grid enthusiasts need to familiarize themselves with the particular zoning restrictions that apply to the selected area to ensure that their intended activities—building off-grid homes or establishing agricultural systems, for example—comply with local ordinances. Infractions of these rules may result in fines, legal issues, and even the demolition of buildings in violation.

Another essential part of the legal aspects of living off the grid is building codes. The criteria and requirements for building projects are outlined in these regulations, created by municipal authorities to guarantee environmental compliance, structural integrity, and safety. Off-grid buildings, which are frequently distinctive and unusual, could be criticized if they depart from accepted building practices. Obtaining the required permissions and following building codes are crucial to prevent legal issues. Comprehensive plans and specifications must be submitted for approval to complete this procedure and demonstrate compliance with safety standards and environmental regulations. Off-grid residents protect the endurance and stability of their built homes and follow the law by acquiring the necessary permissions.

Managing waste is an essential part of living off the grid, and there are legal implications for handling and disposing of waste. Both organic and non-organic waste disposal may be governed by local rules, requiring adherence to particular protocols. Off-grid aficionados frequently use eco-friendly garbage disposal techniques, recycling programs, and composting systems. Complying with local waste management laws guarantees that off-grinders protect the environment while avoiding possible legal consequences for inappropriate garbage disposal.

Water rights and usage restrictions are crucial legal factors for people who want to live off the grid, especially in places where water sources are pooled or subject to laws. Legal restrictions may apply to water availability for livestock, irrigation, and residential usage. To avoid disputes with nearby landowners and to guarantee local regulations are followed, those who live off the grid should thoroughly investigate and comprehend the water rights in the area they have selected. Permits may occasionally be needed to extract or divert water, highlighting the significance of taking a proactive and knowledgeable approach to water-related legal issues.

Legal issues are equally relevant when it comes to renewable energy. Permitting and regulatory compliance may be necessary to install solar panels, wind turbines, or other alternative energy systems. For example, local authorities may need to approve grid-tied or off-grid solar projects to guarantee safety and compliance with electrical rules. Off-grid enthusiasts who secure the necessary permissions adhere to the law and help their communities accept and integrate renewable energy methods more widely.

Moreover, growing crops and rearing livestock are common aspects of off-grid living and may be governed by agricultural laws. Zoning laws may regulate the allowable size of farming activities, and adherence to these rules is necessary to avoid legal trouble. Organic agricultural methods can also call for certification and conformity to particular guidelines if used. The sustainable and lawful pursuit of off-grid farming is aided by in-depth study and comprehension of local agricultural laws.

Off-grid people should be aware of the legal ramifications surrounding easements, which permit using or accessing another person's property for a particular purpose. The development and shape of the site may be impacted by existing easements, which may restrict certain activities or dictate where structures can be erected. Any easements on the property must be well investigated and understood to prevent future conflicts and legal difficulties.

Even while the legal factors covered up to this point are crucial for executing off-grid initiatives, it's also necessary to acknowledge that different jurisdictions have different legal frameworks.

Off-grid enthusiasts interested in learning more about the restrictions that apply to their chosen place should visit local authorities or seek legal guidance. This proactive strategy assists people in making educated decisions, navigating complex legal issues, and establishing a sustainable and legal off-grid lifestyle.

To summarize, permits and legal issues are essential to the off-grid living adventure. Off-grid enthusiasts must be proactive and watchful to ensure compliance with local laws, from managing waste, water rights, and agricultural restrictions to navigating land-use regulations and acquiring building permissions. Although the appeal of living alone frequently centers on autonomy, understanding and observing legal frameworks is necessary for a smooth and successful assimilation into the larger community. By adopting a thorough awareness of the legal environment, people can go off the grid confidently, knowing that their activities comply with the law and their personal goals.

CHAPTER II

Energy Independence

Solar power systems for homesteads

In the quest for sustainable and off-grid living, harnessing solar power is a cornerstone technology, offering homesteads a reliable and environmentally friendly energy source. Integrating solar power systems on homesteads represents a shift toward self-sufficiency and reduced reliance on traditional grid-based electricity. This section explores the various aspects of solar power systems for homesteads, encompassing the technology behind solar panels, the components of a solar power system, the advantages and challenges of solar energy, and considerations for successful implementation.

At the heart of solar power systems are photovoltaic (PV) cells, which convert sunlight into electricity. These cells are typically made from semiconductor materials such as silicon. When sunlight strikes the PV cells, it excites electrons, generating a flow of direct current (DC) electricity. This DC electricity is converted into alternating current (AC) by an inverter, making it compatible with the standard electrical appliances used in homes. The modular nature of PV cells allows for the creation of solar panels, which can be installed on rooftops, ground- mounted structures, or integrated into building materials, providing flexibility in their deployment on homesteads.

A typical solar power system for a homestead comprises several vital components. Solar panels, as mentioned, are the primary electricity generators and are usually grouped in an array. These panels are connected to inverters, which play a crucial role in converting the DC electricity produced by the panels into AC electricity for household use.

Homesteads often incorporate battery storage systems to store excess energy for later use. Batteries, such as lithium-ion or lead-acid batteries, store the surplus electricity generated during sunny periods and release it when sunlight is insufficient, ensuring a continuous and reliable power supply. Additionally, solar power systems may include charge controllers, which regulate the charging of batteries and prevent overcharging, extending the lifespan of the batteries.

The advantages of solar power systems for homesteads are manifold. Foremost among these is the generation of clean and renewable energy. Solar power is environmentally friendly, producing electricity without emitting greenhouse gases or other pollutants associated with conventional energy sources. This characteristic aligns seamlessly with the ethos of sustainable living, contributing to reduced carbon footprints and mitigating the impact of climate change. Moreover, solar power systems provide energy independence to homesteads, freeing them from dependence on centralized utility grids. This autonomy is particularly beneficial in remote or rural areas where access to grid-based electricity may be limited. The potential for financial savings is also significant, as solar power systems can lead to reduced or eliminated electricity bills over the long term, offering a return on the initial investment.

However, despite the numerous advantages, solar power systems for homesteads also present challenges. The initial cost of purchasing and installing solar panels and associated components can be a barrier for some individuals. While the prices have decreased over the years, the upfront investment remains a consideration for those contemplating a shift to solar energy. Additionally, the intermittent nature of sunlight poses a challenge to consistent energy production. Weather conditions, seasonal variations, and nighttime hours impact the sunlight available for solar panels to generate electricity. To address this, homesteads often integrate battery storage systems to store excess energy during peak sunlight hours for use during periods of low sunlight.

Successful implementation of solar power systems on homesteads requires careful consideration of various factors. One critical consideration is the orientation and tilt of the solar panels. To maximize energy production, panels should ideally face south in the northern hemisphere or north in the southern hemisphere. The tilt angle should be optimized based on the geographical location to ensure that panels receive the most sunlight throughout the day. Shading from trees, buildings, or other obstructions should be minimized to prevent loss of efficiency. Additionally, regular maintenance and cleaning of solar panels are essential to ensure optimal performance. Dust, debris, and bird droppings can reduce panels' efficiency, so periodic cleaning is recommended to maximize energy output.

Integration with the existing electrical system of a homestead also requires careful planning. The inverter, which converts DC to AC, should be appropriately sized to handle the household's electricity demand. Battery storage systems need to be sized to meet the energy needs during periods of low sunlight. Conducting a thorough energy audit to understand the electricity consumption patterns and determine the capacity required for solar panels and storage systems is essential. Seeking professional guidance during the design and installation phases can help optimize the system for the specific needs of the homestead.

In conclusion, homestead solar power systems represent a robust and sustainable solution for achieving energy independence and reducing environmental impact. The technology behind solar panels, coupled with the various components of a solar power system, enables homesteads to generate clean electricity from sunlight. While the initial investment and intermittent sunlight present challenges, the long-term benefits, including financial savings and reduced reliance on traditional grids, make solar power systems attractive for those committed to off-grid and sustainable living. Successful implementation requires careful planning, consideration of factors such as panel orientation system sizing, and ongoing maintenance. As technology advances and costs decline, solar power systems are poised to play an increasingly integral role in shaping the future of off-grid homesteads and contributing to a more sustainable and resilient way of life.

Wind energy options

As the global pursuit of sustainable and renewable energy intensifies, wind energy emerges as a prominent and viable option for powering homes, businesses, and entire communities. Harnessing the wind's kinetic energy, wind energy options have evolved to encompass a range of technologies, from traditional windmills to modern wind turbines. This section explores the diverse aspects of wind energy options, delving into the underlying technology, the different types of wind turbines, advantages and challenges, and considerations for successful integration into various settings.

At the core of wind energy technology is converting the kinetic energy of moving air into electrical power. Wind turbines, the primary devices for capturing wind energy, are engineered to achieve this transformation efficiently. The essential components of a wind turbine include the rotor, which consists of two or more blades that capture the wind, a hub that connects the blades to the main shaft, and a generator that converts the rotational energy into electricity. Modern wind turbines often feature advanced technologies, such as pitch control and yaw systems, to optimize energy capture and ensure operational efficiency.

Wind turbines come in various types, with horizontal-axis and vertical-axis turbines being the two main categories. Horizontal-axis turbines are the most common and recognizable, featuring a three-blade design mounted atop a tall tower. These turbines are designed to face the wind and capture wind energy across various wind speeds. In contrast, vertical-axis turbines have a rotor that spins around a vertical axis, resembling an eggbeater or helix. While less common, vertical-axis turbines offer advantages in terms of lower wind requirements for operation and simplified maintenance due to ground-level accessibility.

Advancements in wind energy technology have led to the development of onshore and offshore wind farms. Onshore wind farms are situated on land, often in areas with consistently strong and predictable winds. These farms can range from small installations to large-scale projects comprising multiple turbines. On the other hand, offshore wind farms are located in bodies of water, typically along coastlines. Offshore locations often experience more robust and consistent winds, making them ideal for harnessing wind energy. The turbines in offshore wind farms are specifically designed to withstand the corrosive effects of saltwater and challenging marine conditions.

The advantages of wind energy are substantial and contribute to its growing popularity as a renewable energy option. Firstly, wind is an abundant and inexhaustible resource, making wind energy a sustainable and environmentally friendly solution. Unlike fossil fuels, wind power generation produces no greenhouse gas emissions, contributing to efforts to mitigate climate change. Wind turbines have a relatively low environmental impact compared to traditional energy sources, and their land footprint allows for other land uses, such as agriculture, to coexist. Furthermore, wind energy projects can provide economic benefits by creating manufacturing, installation, and maintenance jobs, contributing to local economies.

However, wind energy also poses specific challenges that must be addressed for successful implementation. One of the primary challenges is variability in wind speed and direction. The intermittent nature of wind requires additional energy storage or backup systems to ensure a continuous power supply. Battery storage technologies, advancements in grid management, and complementary energy sources, such as solar power or hydropower, can help mitigate the impact of this variability. Additionally, wind turbines' visual and noise impact can be a concern, especially in residential areas. Proper siting and community engagement are crucial to addressing these concerns and ensuring that local communities accept and support wind energy projects.

Successful integration of wind energy options into various settings requires careful consideration of several factors. Site selection is paramount, with wind assessments conducted to determine the wind resource potential of a particular location. Areas with consistent and strong winds are ideal for maximizing energy production. Environmental impact assessments should also be conducted to evaluate potential effects on wildlife, landscapes, and local ecosystems. Engaging with local communities throughout the planning and implementation phases fosters transparency and helps address any concerns or opposition.

Furthermore, regulatory and policy considerations are crucial in deploying wind energy projects. Governments and local authorities may offer incentives, subsidies, or streamlined permitting processes to encourage wind energy development. Understanding and complying with regulations related to land use, environmental impact, and safety standards are essential for successfully approving and implementing wind energy options.

Offshore wind energy, in particular, presents unique considerations. The installation and maintenance of offshore wind farms require specialized technologies and expertise. The distance from shore, harsh marine conditions, and the need for robust foundations demand careful planning and engineering. However, the potential for higher wind speeds and reduced visual impact make offshore wind a promising frontier for expanding wind energy capacity.

In conclusion, wind energy options represent a compelling and increasingly significant component of the global transition to sustainable and renewable energy sources. The technology behind wind turbines has evolved, offering a range of options for harnessing wind power, from onshore and offshore wind farms to small-scale installations. The advantages of wind energy, including its sustainability, minimal environmental impact, and economic benefits, position it as a crucial player in the quest for clean energy. While intermittency and community concerns exist, ongoing technological advancements and comprehensive planning can address these issues. As the world continues prioritizing the shift to renewable energy, wind power is a formidable and promising solution, contributing to a more sustainable and resilient energy future.

Micro-hydro systems

In pursuing sustainable and off-grid living, micro-hydro systems emerge as a versatile and environmentally friendly option for harnessing the power of flowing water to generate electricity. Micro-hydro systems represent a scaled-down version of traditional hydropower systems, offering a decentralized and accessible solution for individuals, communities, and homesteads seeking to produce clean energy. This section explores the various facets of micro-hydro systems, covering the underlying technology, the components of a micro-hydro system, advantages and challenges, and considerations for successful implementation.

Micro-hydro systems operate on the fundamental principle of converting the kinetic energy of flowing water into mechanical energy, which is then transformed into electrical power. The essential components of a micro-hydro system include a water source, such as a stream or river, a diversion structure to direct water flow, a penstock (pipeline) to convey water to a turbine, a turbine itself, a generator, and an electrical control system. The flowing water is channeled through the penstock to turn the turbine blades, causing the turbine to rotate. This rotational motion is then transferred to the generator, converting the mechanical energy into electrical energy. The generated electricity can be used immediately or stored in batteries for later use, providing a reliable and sustainable power source.

Micro-hydro systems come in various configurations, including high-head and low-head systems, each suited to different types of water resources. High-head systems utilize steep slopes and fast-flowing water to generate electricity efficiently, while low-head systems are designed for slower-flowing water and less elevation change. The choice between these configurations depends on the water source's specific characteristics and the installation site's terrain.

One of the critical advantages of micro-hydro systems lies in their ability to provide a consistent and reliable source of electricity. Unlike solar and wind power, which are subject to intermittency based on weather conditions, micro-hydro systems can operate continuously if there is a steady water flow. This characteristic makes them particularly well-suited for off-grid and remote locations where other renewable energy sources may be less reliable.

Furthermore, micro-hydro systems have minimal environmental impact when designed and implemented responsibly. They do not emit greenhouse gases during operation, contributing to a reduction in carbon footprints. Micro-hydro projects' ecological footprint is generally smaller than large-scale hydropower installations, with less disruption to ecosystems and aquatic habitats. By harnessing the energy of flowing water in a controlled manner, micro-hydro systems can coexist harmoniously with the natural environment.

However, challenges and considerations exist in the implementation of micro-hydro systems. One significant consideration is the availability and reliability of the water source. A consistent and adequate flow of water is essential for the reliable operation of a micro-hydro system. Seasonal variations, precipitation changes, or watercourse alterations can impact the system's performance. Detailed site assessments, hydrological studies, and monitoring of the water source are crucial to ensure the feasibility and sustainability of micro-hydro projects.

Another consideration is the potential impact on aquatic ecosystems and water quality. Careful design and implementation are necessary to prevent adverse effects on fish habitats, marine organisms, and water quality. Fish-friendly turbine designs, fish passages, and sediment control measures can be incorporated to mitigate environmental impacts. Additionally, regulatory approvals and compliance with environmental standards are essential to ensure that micro-hydro projects adhere to legal requirements and best practices.

The scale of micro-hydro systems also influences their cost-effectiveness and applicability. While micro-hydro systems are generally more affordable than larger hydropower projects, the economics of a specific installation depend on factors such as the available water flow, head (drop in elevation), and the local cost of equipment and installation. The feasibility of micro-hydro systems is often optimized in situations where the terrain and water resources align favorably with the system's characteristics.

Successful implementation of micro-hydro systems involves comprehensive planning and system design. Site selection is a critical first step, considering factors such as the availability of a reliable water source, topography, and environmental considerations. Detailed feasibility studies inform the design process, including hydrological assessments and environmental impact analyses. The selection of appropriate equipment, including turbines and generators, is tailored to the site's specific conditions.

Community engagement is vital during the planning and implementation phases of micro-hydro projects. Involving local communities in decision-making processes, addressing concerns, and ensuring that the project's benefits are shared with the community contribute to micro-hydro systems' long-term success and acceptance. Additionally, knowledge transfer and capacity building within local communities empower them to take an active role in the operation and maintenance of the system.

Moreover, governmental policies and regulations play a crucial role in developing micro-hydro projects. Supportive policies, incentives, and regulatory frameworks that encourage the development of small- scale hydropower can facilitate the widespread adoption of micro-hydro systems. Clear guidelines on permitting, environmental compliance, and safety standards contribute to a conducive environment for deploying micro-hydro projects.

In conclusion, micro-hydro systems present a viable and sustainable option for generating electricity, particularly in off-grid and remote locations. The technology behind micro-hydro systems, focusing on harnessing the energy of flowing water, offers a reliable and continuous power source. Advantages include minimal environmental impact, consistent energy production, and the potential for community empowerment. However, challenges related to water source variability, ecological considerations, and economic viability necessitate careful planning and thorough assessments. With proper site selection, community engagement, and adherence to regulatory standards, micro-hydro systems can contribute significantly to the global transition toward clean and renewable energy sources, providing a reliable and environmentally friendly solution for decentralized power generation.

Efficient use of energy in off-grid living

In the realm of off-grid living, where the quest for self-sufficiency meets the imperative of environmental sustainability, the efficient use of energy emerges as a linchpin for success. Disrupting from traditional utility grids and embracing an off-grid lifestyle necessitates a thoughtful approach to energy consumption, conservation, and generation. This section delves into the various facets of efficient energy use in off-grid living, encompassing energy conservation principles, optimizing energy systems, integrating intelligent technologies, and fostering a mindset shift towards responsible energy practices.

At the core of efficient energy use in off-grid living lies the principle of energy conservation. This involves minimizing energy wastage and maximizing the output derived from the available resources. One of the fundamental strategies in energy conservation is adopting energy- efficient appliances and lighting. LED bulbs, for instance, consume significantly less energy than traditional incandescent bulbs and have a longer lifespan. Energy- efficient appliances, ranging from refrigerators to water heaters, are designed to perform optimally while minimizing energy consumption. Choosing appliances with high energy efficiency ratings, such as ENERGY STAR-certified products, becomes a pivotal consideration for off-grid enthusiasts seeking to reduce their overall energy demand.

The efficient design and insulation of living spaces are crucial for energy conservation in off-grid living. Well-insulated homes retain heat during cold periods and remain calm in warmer weather, reducing the need for additional heating or cooling systems. Passive solar design principles, such as strategically positioning windows to capture sunlight, can contribute to natural heating and lighting, minimizing reliance on energy- consuming solutions. Energy-efficient building materials and construction techniques, coupled with thoughtful design choices, create homes that are not only environmentally responsible but also conducive to a comfortable and sustainable off-grid lifestyle.

Optimizing energy systems is another critical element of efficient energy use in off-grid living. This involves choosing and configuring energy generation systems, such as solar panels, wind turbines, or micro-hydro systems, to match the specific energy needs of the household. Conducting a thorough energy audit provides insights into energy consumption patterns, allowing off-grid enthusiasts to right-size their energy systems. Proper system sizing ensures that the energy generated meets the demand without excess, minimizing waste and enhancing overall efficiency.

Incorporating energy storage solutions, such as batteries, is integral to optimizing energy systems in off-grid living. Energy storage allows for capturing and storing excess energy generated during peak production periods, making it available for use during low or no energy generation. This addresses the intermittency of renewable energy sources, such as solar or wind, and ensures a continuous and reliable power supply. Advances in battery technology, including lithium-ion batteries, have enhanced off-grid energy systems' efficiency and storage capacity, contributing to the overall resilience of off-grid living.

The integration of intelligent technologies further enhances the efficiency of energy use in off-grid living. Smart home systems, equipped with sensors and automation, enable the monitoring and controlling of energy-consuming devices. This includes adjusting lighting, heating, and cooling systems based on occupancy and preferences. Smart thermostats, for example, learn from user behavior and optimize heating or cooling schedules to minimize energy consumption. Additionally, energy monitoring systems provide real-time insights into energy usage patterns, empowering off-grid residents to identify and address sources of inefficiency.

The adoption of energy management practices complements energy-efficient appliances. Simple habits, such as turning off lights when not in use, unplugging chargers and electronics, and minimizing phantom loads, reduce overall energy consumption. The cultivation of energy-conscious behavior becomes a cornerstone of efficient energy use in off-grid living, emphasizing the importance of mindfulness in daily energy-related activities.

Renewable energy sources like solar and wind are central to the off-grid energy paradigm. Efficient use of these resources involves optimizing their capture and conversion to electricity. Solar panels, for instance, should be strategically positioned to receive maximum sunlight exposure throughout the day. Wind turbines benefit from proper siting in areas with consistent and sufficient wind speeds. The combination of solar and wind systems, sometimes complemented by micro-hydro systems, creates a diversified and reliable renewable energy portfolio for off-grid living.

In addition to generating electricity, off-grid living often involves directly using renewable energy for heating and cooking. Solar water heaters, for example, utilize sunlight to heat water for domestic use, reducing the need for conventional water heating methods. Solar cookers harness solar energy to cook food, offering an eco- friendly alternative to traditional stoves. These direct-use applications of renewable energy contribute to overall efficiency by minimizing the conversion losses associated with generating electricity.

Adopting energy-efficient transportation solutions is an essential component of efficient energy use in off-grid living. Electric vehicles (EVs) powered by renewable energy sources align with the ethos of sustainability and contribute to reducing a household's carbon footprint.

Charging EVs using on-site renewable energy generation ensures that the environmental benefits extend beyond the confines of the home. Additionally, integrating energy-efficient transportation aligns with the broader goal of achieving a holistic and sustainable off-grid lifestyle.

Efficient energy use in off-grid living is a technical endeavor and a mindset shift. It involves cultivating an awareness of energy consumption patterns, understanding the environmental impact of energy choices, and making intentional decisions to minimize waste. Education and awareness-building within off-grid communities are crucial in fostering this mindset shift. By emphasizing the interconnectedness of individual actions with broader environmental sustainability goals, off-grid enthusiasts can contribute to a collective consciousness that values and prioritizes responsible energy practices.

In conclusion, efficient energy use in off-grid living represents a multifaceted approach that spans technology, design, behavior, and awareness. Energy conservation principles guide choices in appliances, lighting, and building design while optimizing energy systems to ensure that generation meets demand efficiently. Intelligent technologies and renewable energy sources further enhance the efficiency of off-grid living, offering resilience and sustainability. Cultivating energy-conscious behavior and fostering a mindset shift toward responsible energy practices contribute to a holistic approach that aligns with the principles of off-grid living. As the global focus on sustainable living intensifies, efficient energy use is a foundational element for those embracing the transformative journey of off-grid living.

CHAPTER III

Water Management

Rainwater harvesting systems

In pursuing sustainable and self-sufficient living, rainwater harvesting systems emerge as a practical and environmentally conscious solution to water scarcity. These systems harness the power of nature, capturing and storing rainwater for various domestic and agricultural uses. Rainwater harvesting is a time-tested method that has advanced with technology to provide a dependable, decentralized water source for people, communities, and homesteads. The many aspects of rainwater harvesting systems are examined in this section, along with the underlying technology, system elements, advantages and disadvantages, and implementation best practices.

The foundation of rainwater harvesting systems is the straightforward but brilliant idea of collecting rainwater as it falls and channeling it toward storage for later use. A conveyance system, distribution system, storage tanks, and catchment surface are the fundamental parts of a rainwater harvesting system. The catchment surface, usually a building's roof, collects rainwater. The rainwater collection is directed to storage tanks by the conveyance system, which consists of gutters and downspouts. The collected rainwater is kept in these storage tanks, constructed of metal, plastic, or concrete until it is required for different purposes. A distribution system then directs the collected rainwater to various locations, including gardens, toilets, and, with the proper preparation, drinking water sources.

Through technological advancements, rainwater harvesting devices have become more effective and user-friendly. For example, first flush diverters reroute the initial, potentially contaminated runoff from the catchment surface away from the storage tank. By doing this, the storage system is guaranteed to receive only cleaner rainwater. Filtration systems improve the quality of captured rainwater for various applications by removing debris, silt, and pollutants. They can be as basic as mesh filters or as complex as multi-stage filters. Furthermore, techniques for treating water, including UV disinfection or chlorination, can be used to ensure that the collected rainwater satisfies safety requirements for drinking.

Rainwater harvesting systems have many advantages and support sustainable water management. Decreased dependency on conventional water sources, like wells or municipal water supply, is one of the main benefits. In some situations, rainwater harvesting can replace or augment traditional water sources as a decentralized, alternate water source. This improves water resilience while also aiding in the preservation of the water resources that are currently available, particularly in areas where water availability is erratic or scarce.

By collecting rainfall that would otherwise run off and be lost to the environment, rainwater harvesting helps to conserve water. This lessens the effects of impervious surfaces and urbanization, frequently leading to increased runoff and decreased groundwater recharge. People and communities can actively contribute to the sustainability of water supplies and the preservation of nearby water habitats by collecting and using rainwater.

Rainwater harvesting systems also encourage a decrease in stormwater runoff, which can contaminate and pollute natural water bodies. These systems contribute to improving water quality and reducing water pollution by collecting precipitation before it becomes runoff. This becomes more important in cities because industrial processes, impervious surfaces, and vehicle runoff contribute to water contamination.

Rainwater harvesting has many residential and agricultural uses because of its adaptability. Rainwater collected from rooftops can be utilized for non-potable indoor uses like toilet flushing and laundry, irrigation, gardening, and landscape upkeep. Rainwater can even be made drinkable with the proper treatment, providing a decentralized and sustainable source of drinking water. This adaptability highlights how rainwater harvesting systems can be used to cover a range of water needs in various settings.

Rainwater harvesting systems require careful planning and execution to be implemented successfully despite their many advantages. The system's size and design to meet the individual water needs of the user are essential factors to consider. When choosing the suitable storage capacity and system components, considerations include catchment area, rainfall patterns, and the anticipated applications of the harvested rainwater. To build a system that collects and stores rainwater as efficiently as possible, one must thoroughly grasp the local environment, including average rainfall and seasonal changes.

Another factor that affects rainwater harvesting systems' long-term efficacy is maintenance. Cleaning gutters, downspouts, and filters regularly are required to avoid silt and debris accumulation, which can lower the quality of rainfall collected. Storage tanks should have regular sanitizations and leak inspections to guarantee water quality. Paying attention to maintenance procedures extends the system's life and ensures the dependability of collected rainwater for various uses.

Regarding potable usage, water quality is a crucial component of rainwater gathering. Rainwater harvesting may get contaminated by pollutants, leaves, or bird droppings from the catchment surface. Sulfur filters, first flush diverters, and other filtering devices are essential for raising water quality. Furthermore, routine testing and observation of the collected rainwater can reveal details about its chemical makeup and recommend suitable treatment methods for drinking.

Rainwater harvesting system deployment is also influenced by local legislation and regulatory issues. Collected rainwater, particularly for drinkable uses, may be subject to strict laws in some areas. Ensuring compliance with legal obligations necessitates obtaining relevant permits and following local norms. The seamless integration of rainwater harvesting systems into communities and homesteads can be facilitated by interacting with local authorities and requesting information on regulatory matters.

An additional factor that needs consideration is how rainwater harvesting systems are seen in society and culture. In certain areas, rainwater harvesting customs may already be established, and people may be open to adopting new technology. In other situations, raising awareness and educating the public could be essential to promoting the advantages of rainwater harvesting and clearing up any misunderstandings. Rainwater collecting techniques are successfully adopted when the community is involved and consulted, which instills a sense of accountability and ownership in the locals.

In summary, rainwater collecting systems provide a decentralized, adaptable water source for various uses, making them a realistic and sustainable response to the water shortage. Rainwater harvesting is now feasible and flexible for multiple situations because of the technology underlying these systems and advancements in filtration and treatment. Rainwater harvesting is an integral part of sustainable water management because of its advantages, which include decreased dependency on conventional water sources, preservation of water resources, and improved water quality. Careful planning, community engagement, and adherence to best practices can effectively install rainwater harvesting systems, even in the face of obstacles relating to system design, maintenance, water quality, and legal issues. Rainwater harvesting provides resiliency and self-sufficiency in pursuing a water-secure future as the globe struggles with water scarcity and the necessity of sustainable water use.

Sustainable good usage

Wells have been integral to human civilizations for millennia, providing a dependable water source for various domestic, agricultural, and industrial needs. In sustainable living, using wells adds significance as communities and individuals seek to balance their water needs with preserving groundwater resources. While meeting water demands, sustainable, healthy usage ensures the durability and health of aquifers through careful practices, technologies, and management strategies. This section examines the various facets of sustainable beneficial usage, covering water-saving techniques, effective pumping methods, aquifer management, and the value of community involvement.

The prudent management of aquifers, the underground geological formations that store and transport groundwater, is essential to the sustainable use of wells. Aquifers are naturally occurring reservoirs that can hold enormous volumes of water that can be drawn out via wells. On the other hand, excessive groundwater extraction can result in aquifer depletion, which can have long-term effects like decreased water availability and land subsidence. To maintain sustainable management, the extraction rate must be monitored and controlled to ensure it stays within the aquifer's natural replenishment rate. To minimize overexploitation and protect this vital water resource, sustainable aquifer management must include implementing well-spacing rules and the evaluation of an aquifer's safe yield.

An essential component of sustainable healthy usage is efficient pumping methods. Wells was fitted with primary pumps frequently driven by wind or human labor. Modern well pumps, on the other hand, provide more efficiency and control. These include submersible and jet pumps that run on electricity or solar power. Pump wear and energy consumption can be minimized by adjusting pump speed in response to water demand using variable frequency drives (VFDs). These technical developments improve the sustainability of good usage by reducing the environmental effect of energy consumption and streamlining the extraction process.

Sustainable healthy utilization depends on water conservation techniques, impacting personal behavior and agricultural activities. Simple household actions like leak repairs, installing water-efficient equipment, and adopting conscientious water usage practices all help overall conservation. Precision irrigation, soil moisture monitoring, and drought-tolerant crop selection can all help with agricultural irrigation, which uses a lot of groundwater. To minimize the need for excessive groundwater pumping, sustainable agriculture strongly emphasizes increasing soil health, cutting down on water waste, and enhancing water usage efficiency.

Participation in the community is essential to sustainable well-being. Involving the local community in healthy management promotes a sense of accountability and ownership. Participatory methods, such as community-based monitoring and decision-making, enable locals to tackle water-related issues and implement long-term fixes. A foundation for educated decision-making and the long-term protection of healthy resources is laid by educating communities about the value of groundwater, the adverse effects of over-extraction, and the advantages of conservation efforts.

An essential component of sustainable healthy usage is groundwater recharge, ensuring that surface water sources and precipitation refill aquifers. The sustainability of beneficial resources is enhanced by applying techniques to improve recharge, such as building artificial recharge basins or safeguarding natural recharge sites. Moreover, groundwater recharging can be enhanced via rainwater harvesting and stormwater management techniques, which might lessen the need for wells during high demand.

Achieving sustainable well usage also requires addressing the quality of the water. Groundwater quality can be compromised by contaminants that seep in from inadequately managed wastewater, industrial processes, and agricultural runoff. Groundwater integrity is protected by putting protective measures in place, including buffer zones between farming areas and wells, and by enacting laws that forbid the discharge of hazardous materials close to healthy sites. It's crucial to regularly test and monitor well water for pollutants such as bacteria, nitrates, and other impurities to ensure the extracted water satisfies safety and health requirements.

With altered precipitation patterns and more weather unpredictability impacting groundwater recharge, climate change poses new difficulties for sustainable, healthy utilization. An example of sustainable adaptation techniques is assessing a well's susceptibility to shifting weather patterns, conserving water during dry spells, and creating backup plans for instances of water scarcity. Groundwater resources are protected against climate change by incorporating climate resilience into healthy management techniques.

To summarize, sustainable healthy usage is a dynamic and all-encompassing strategy that includes groundwater recharge, community involvement, effective pumping technology, aquifer management, water conservation measures, and climate resilience. It necessitates a paradigm change toward prudent water management, balancing the need to protect aquifers for future generations and groundwater exploitation. Adopting sustainable, healthy usage methods becomes crucial as the world's water demand rises, particularly in light of population increases and shifting climate conditions. The amalgamation of technological innovations, community involvement, and ecological preservation strategies enhances the robustness and endurance of healthy supplies, guaranteeing a future fair and sustainable water supply.

Water purification techniques

Access to clean and safe drinking water is a fundamental human right and a critical component of public health. However, various environmental factors, industrial activities, and population growth contribute to water pollution, necessitating effective water purification techniques. These methods are essential for guaranteeing that water sources are free of pollutants, pathogens, and impurities and adhere to safety and health regulations. This section examines many approaches to water purification, ranging from conventional processes to cutting-edge technology, emphasizing their foundations, uses, and importance in the fight for clean water worldwide.

Boiling is one of the oldest and most popular ways of purifying water. Boiling water destroys viruses, germs, and parasites, ensuring drinking is safe. This approach is beneficial in environments with limited resources because it may only sometimes be possible to use advanced purification technology. Boiling gets rid of microorganisms, but it doesn't get rid of chemicals or suspended particles. Additionally, it needs a dependable energy supply, which might not be easily accessible in some areas.

Filtration and sedimentation are two other tried-and-true techniques. Larger particles settle more quickly in this process because the water is left undisturbed. A filter subsequently collects the residual suspended particles after the water has become more apparent. Sedimentation and filtering work well to remove visual contaminants and turbidity, but they might not be able to remove chemical pollutants or microscopic microorganisms. However, these techniques offer a valuable and affordable way to enhance water clarity in various settings.

Adding chlorine compounds to water is a common chemical disinfection technique called chlorination. Since chlorine efficiently eliminates germs and viruses, it is preferred for extensive water treatment in municipal water systems. However, research for substitute disinfection techniques has been spurred by worries about developing disinfection byproducts and the flavor and odor of chlorinated water. Furthermore, chlorination may not eradicate all parasites and chemical pollutants.

UV irradiation is a cutting-edge technology that has become more well-known in recent years for purifying water. UV light damages microorganisms' DNA, stopping them from increasing. UV purification offers a chemical-free, eco-friendly option and is very effective against bacteria, viruses, and protozoa. This technique, which guarantees that water is safe for consumption without changing its taste or odor, is frequently employed in point-of-use water treatment systems. Water turbidity may affect the efficacy of UV treatment, and it may not be able to eliminate chemical pollutants or particles.

Various pollutants can be effectively and widely removed from water using activated carbon filtration. Because of its porous nature, activated carbon can draw in and adsorb multiple contaminants, such as chemicals, certain metals, and organic molecules. This process effectively eliminates chlorine and its byproducts, improving taste and odor. However, because activated carbon has a limited capacity, it must be regularly replaced or regenerated to continue performing at its best. Additionally, activated carbon filtration might not completely eliminate some minerals, salts, or microbiological pollutants.

Reverse osmosis (RO) is a membrane-based method of purifying water that successfully removes a wide range of pollutants by forcing water through a semi-permeable membrane under pressure. RO is very effective When getting rid of bacteria, heavy metals, dissolved salts, and other contaminants. It is frequently utilized in large-scale desalination plants and residential water treatment systems. RO has disadvantages despite its efficacy, such as high energy consumption and concentrated brine waste output. Furthermore, the procedure can eliminate vital minerals from the water, requiring further mineralization.

A resin and the water exchange ions are part of the ion exchange water treatment process. This method is very good at softening water by eliminating the calcium and magnesium ions that cause water to be complicated. To stop scale accumulation in pipes and appliances, ion exchange is frequently utilized in home water softeners. Nevertheless, it might be unable to deal with other kinds of pollutants, and chemical solutions are needed for the ion exchange resins to regenerate.

Flocculation and coagulation are crucial processes for eliminating suspended particles when treating water. Coagulation entails adding chemical coagulants, such as ferric chloride or aluminum sulfate, to destabilize the particles in the water. The next step is flocculation, in which light swirling or mixing encourages the development of more extensive, aggregated particles (flocs). The sediment and the purified water can be separated once these flocs settle. While flocculation and coagulation are good at lowering turbidity and particle matter, more is needed for microbiological eradication with further disinfection procedures.

The term "advanced oxidation processes" (AOPs) refers to methods for separating organic and inorganic pollutants in water using solid oxidants. Ozonation, hydrogen peroxide therapy, and UV-based advanced oxidation are a few AOPs. These reactions produce Highly reactive hydroxyl radicals, which convert contaminants into innocuous byproducts. AOPs work well against various pollutants, including medications and newly discovered pollutants. Nevertheless, the expense, energy needs, and possible byproduct generation may restrict their use.

To sum up, several water purification methods can be employed to tackle the intricate problem of guaranteeing that people have access to uncontaminated and safe drinking water. Every methodology has advantages and considerations, ranging from conventional methods like boiling and sedimentation to cutting-edge technologies like reverse osmosis and sophisticated oxidation processes. The kind of contaminants, the extent of the water treatment, the availability of energy, and the intended level of water quality all have a role in choosing a suitable purification technique. Research and development in water purification technologies will be essential to meeting the challenge of the growing global demand for safe water and ensuring the health and welfare of communities everywhere.

Smart water usage practices

In an era marked by growing environmental consciousness and a deepening awareness of resource scarcity, intelligent water usage practices have emerged as a critical component of sustainable living. As the global population expands and climate change introduces uncertainties to water availability, adopting competent and efficient approaches to water consumption becomes imperative. This section examines a range of innovative technology solutions and responsible water usage practices, emphasizing their importance in reducing water stress, protecting ecosystems, and promoting a more sustainable connection with this essential resource.

Developing awareness and forming water-saving behaviors on a personal level are essential steps toward wise water use. Water waste in homes can be decreased by following easy steps like swiftly addressing leaks, shutting off the faucet when brushing your teeth, and only running dishwashers and washing machines when fully loaded. Installing low-flow showerheads and faucets can significantly reduce water usage without sacrificing functionality. These little but significant adjustments highlight how crucial personal accountability is to maximizing water consumption and reducing waste.

Intelligent water usage techniques heavily depend on landscaping decisions, especially in areas where outdoor water use accounts for a large share of home demand. Irrigation is not as necessary with xeriscaping, a landscaping technique that stresses the use of native and drought-tolerant plants. Mulching helps reduce the impacts of evaporation by keeping the soil moist around plants. Furthermore, using rainwater collecting systems to meet landscaping needs while taking advantage of natural rainfall is consistent with the ideals of sustainability. These landscaping techniques exemplify how deliberate decisions about outdoor water use can support water conservation initiatives.

The emergence of the Internet of Things (IoT) has led to the development of intelligent technologies, which provide creative ways to improve water efficiency across industries. With the help of sensors and integrated weather data, smart irrigation systems optimize watering programs in real-time. These methods guarantee that plants get enough moisture without using extra water. Devices for leak detection and monitoring minimize damage and stop water loss by warning of possible water leaks early. Moreover, intelligent water meters make real-time tracking of users' water consumption possible, which raises awareness and promotes water-wise behavior.

Precision farming techniques are a wise way to use water in agriculture, which uses many of the world's water resources. Farmers can determine the precise amount of water crops require using remote sensing devices and soil moisture sensors. Water loss from evaporation or runoff is reduced using drip irrigation and other water-efficient irrigation techniques, which supply water directly to the root zone. Moreover, maximizing water use efficiency, agroecological techniques, and adopting crop cultivars resistant to water stress support sustainable agriculture.

Creating water-resilient cities in urban design increasingly depends on clever water management techniques. Porous pavements and green roofs are examples of green infrastructure that helps absorb natural water and lessens stormwater runoff, which eases the burden on drainage systems. To ensure a comprehensive approach to urban water sustainability, integrated water management plans consider the entire water cycle, from capture to treatment and reuse. Resilient and effective urban water utilization can be achieved by supporting decentralized water treatment systems and implementing water-sensitive urban design ideas.

Reclaimed water is treated wastewater that satisfies quality standards and is a clever way to supplement water sources. Reclaimed water can be used for non-potable uses such as cooling systems, industrial processes, and landscape irrigation, which lessens the strain on freshwater resources. Using treated wastewater to its fullest potential is a practice that is especially pertinent in areas where water is scarce. It aligns with the concepts of sustainable water management and the circular economy.

Implementing educational programs and community involvement are crucial for wise water use strategies. A culture of responsible water stewardship is fostered by increasing knowledge of the importance of conservation, the worth of water, and the effects of individual and communal activities on water resources.

Incentives for implementing water-efficient technologies combined with community-driven water conservation initiatives foster a sense of shared accountability and empowerment for the sustainable use of water resources.

An essential component of wise water consumption practices is addressing the problem of water contamination. Maintaining water quality is facilitated by avoiding the use of hazardous chemicals, adopting best practices in industry and agriculture, and preventing the contamination of water sources. Wetland conservation and restoration initiatives supplement natural water filtration processes, guaranteeing that water supplies stay clean and appropriate for various applications.

Resilience becomes a critical component of wise water consumption practices in the face of climate change and growing water scarcity. Implementing water reuse technologies, diversifying water sources, and predicting and responding to changing climate conditions are all necessary to build resilient water systems. Projects involving desalination, groundwater recharge, and rainwater harvesting help strengthen the water system's resilience, guaranteeing a steady and sustainable water supply even in the face of climatic uncertainty.

Fostering innovative water usage practices globally requires international cooperation and policy frameworks. We can handle water concerns as a group by exchanging technical advancements, research findings, and best practices. Adopting intelligent water consumption methods across industries and regions is facilitated by policy initiatives that promote sustainable practices, regulate water use, and incentivize water efficiency.

To sum up, intelligent water consumption practices cover various activities, from technological advancements and governmental interventions to individual habits, to maximize water use and guarantee sustainability. Adopting clever and effective strategies for water consumption is crucial as the globe navigates the

complexity of a changing climate, population increase, and environmental deterioration. A comprehensive and progressive approach incorporating mindful habits, technology breakthroughs, sustainable agriculture practices, urban planning initiatives, and community engagement is needed to ensure that future generations can access clean water. Individuals, groups, and countries can help to preserve and practice responsible stewardship of this limited and vital resource by adopting wise water usage practices.

CHAPTER IV

Sustainable Shelter

Eco-friendly construction materials

The construction industry, a powerhouse of economic development and urban expansion, has traditionally been associated with resource-intensive practices and environmental impacts. However, as global awareness of sustainability and environmental conservation has grown, a paradigm has shifted towards eco-friendly construction materials. These materials prioritize environmental responsibility, aiming to reduce the ecological footprint of construction projects and promote a more sustainable built environment. This section explores the diverse landscape of eco-friendly construction materials, from traditional options to cutting-edge innovations. It examines their characteristics, benefits, and contributions to a greener and more sustainable construction sector.

One of the cornerstones of eco-friendly construction is the use of recycled materials. Recycled concrete, for instance, involves repurposing old concrete structures or demolition waste into aggregates for new construction. This not only diverts waste from landfills but also reduces the demand for virgin aggregates, thereby conserving natural resources. Recycled steel, derived from salvaged or scrapped metal, similarly contributes to sustainable construction practices. By incorporating recycled content into structural elements, builders can minimize the environmental impact associated with the extraction and production of raw materials.

Bamboo, a rapidly renewable resource, has gained popularity as a sustainable alternative to traditional hardwoods in construction.

Bamboo grows astonishingly, maturing much faster than conventional trees used for timber. Its strength and versatility make it an ideal material for various applications, from structural elements to flooring and finishes. Moreover, bamboo cultivation enhances soil health and requires minimal chemical inputs, aligning with sustainable agriculture principles.

Engineered wood products, such as laminated veneer lumber (LVL) and cross-laminated timber (CLT), represent sustainable alternatives to traditional timber. These products are manufactured by bonding layers of wood veneers or lumber to enhance structural performance and reduce waste. Engineered wood products provide a durable and resource-efficient solution, utilizing wood from sustainably managed forests and offering a viable substitute for materials with higher environmental impacts.

Rapidly renewable materials extend beyond bamboo to encompass other plant-based options. Hempcrete, a mixture of hemp fibers, lime, and water, offers a lightweight and insulating construction material. Hemp proliferates and requires minimal pesticides, making it an environmentally friendly choice. Similarly, straw bales, an agricultural byproduct, find application in construction as an insulating material. When properly sealed and combined with other elements, straw bale construction provides energy-efficient buildings with a reduced reliance on conventional insulation materials.

Earth-based construction materials, such as adobe, rammed earth, and compressed earth blocks, draw from ancient building traditions and showcase the sustainability inherent in utilizing locally available resources. These materials involve minimal processing and offer excellent thermal mass properties, contributing to building energy efficiency. Using earth-based materials promotes a low-carbon footprint and fosters a connection to local ecosystems and vernacular building techniques.

In the realm of insulation, natural and recycled options are gaining prominence. Sheep's wool, for example, is a biodegradable and renewable insulation material. It possesses excellent thermal properties and does not pose health risks during installation. Cellulose insulation, derived from recycled paper or cardboard, provides an eco-friendly alternative to conventional fiberglass insulation. These materials contribute to energy efficiency by minimizing heat transfer and reducing the need for excessive heating or cooling in buildings.

Innovative technologies have introduced new eco-friendly construction materials to address specific environmental challenges. Self-healing concrete, for instance, incorporates bacteria that produce calcite, helping repair the material's cracks over time. This reduces the need for frequent repairs and enhances the longevity of concrete structures, mitigating the environmental impact associated with maintenance.

Photovoltaic glass, embedded with solar cells, transforms windows and facades into energy-generating surfaces, integrating renewable energy production seamlessly into building design.

Bioplastics, derived from renewable sources such as corn or sugarcane, have found application in construction as an alternative to traditional plastics. Bioplastic materials offer a more sustainable option for various building components, including insulation, cladding, and packaging. These materials can be compostable or biodegradable, addressing concerns related to plastic waste in the construction sector.

Low-impact building systems, such as modular construction and prefabrication, contribute to sustainable construction practices by minimizing waste, optimizing material use, and reducing energy consumption. Modular construction involves assembling building components off-site and transporting them to the construction site for assembly. This method reduces on-site construction time, disturbance to local ecosystems, and waste generation. Additionally, prefabrication allows for precision in manufacturing, leading to more efficient use of materials and resources.

The cradle-to-cradle approach emphasizes the lifecycle impact of materials, encouraging the use of products that can be recycled or safely returned to the environment after use. Cradle-to-cradle certified materials adhere to strict environmental and social criteria, promoting a closed-loop system that minimizes waste and pollution. This approach challenges the traditional linear extraction, production, use, and disposal model, advocating for a circular economy in which materials are continuously regenerated and reused.

While eco-friendly construction materials offer numerous environmental benefits, their adoption faces cost, availability, and industry norms challenges. The upfront cost of some sustainable materials may be higher than conventional counterparts, posing a barrier for budget-conscious projects. However, the long-term benefits, including energy savings, durability, and environmental stewardship, often outweigh the initial investment. The availability of eco-friendly materials may also vary regionally, depending on local industries and regulatory frameworks. Overcoming these challenges requires a concerted effort from the construction industry, policymakers, and consumers to create a supportive environment for the widespread adoption of sustainable building practices.

In conclusion, exploring and integrating eco-friendly construction materials represent a transformative shift towards a more sustainable and responsible built environment. From recycled aggregates and rapidly renewable resources to innovative technologies and low-impact building systems, the diverse array of options available reflects the growing commitment to mitigating the environmental impact of construction activities. As the construction industry continues to evolve, balancing the demand for urban development with environmental stewardship becomes increasingly critical. By embracing and advancing eco-friendly construction materials, stakeholders in the construction sector can contribute to a resilient and sustainable future where buildings harmonize with nature and minimize their ecological footprint.

Passive solar design principles

Passive solar design, rooted in ancient architectural practices and reinvigorated by contemporary sustainability goals, is a testament to the power of harnessing natural energy sources for the built environment. This approach leverages the sun's energy to provide heating, cooling, and lighting for buildings, reducing reliance on mechanical systems and minimizing environmental impact. Passive solar design principles integrate seamlessly into architectural planning, considering the local climate, site orientation, and building materials to optimize energy efficiency. This section explores the fundamental principles of passive solar design, delving into the key strategies architects and designers employ to create buildings that are aesthetically pleasing, environmentally responsible, and energy-efficient.

Orientation plays a pivotal role in passive solar design, with careful consideration given to the positioning of a building about the sun's path. South-facing facades receive the most direct sunlight in the Northern Hemisphere, while north-facing facades are ideal in the Southern Hemisphere. This strategic alignment allows buildings to maximize solar gain during winter when the sun's path is lower in the sky. South-facing windows and walls become solar collectors, absorbing sunlight and converting it into heat that can be stored in thermal mass materials like concrete or masonry. This stored heat is released slowly during the evening, contributing to natural space heating.

Thermal mass is central to passive solar design, serving as a heat storage medium that helps regulate indoor temperatures. Materials with high thermal mass, such as concrete, brick, or stone, absorb and store heat effectively. During the day, when sunlight enters a space, these materials absorb the heat, preventing rapid temperature fluctuations. As the ambient temperature drops in the evening, the stored heat is gradually released, moderating indoor temperatures and enhancing occupant comfort. The strategic placement of thermal mass in areas exposed to sunlight optimizes its effectiveness in capturing and storing solar energy.

Glazing, or using windows and other transparent materials, is critical in passive solar design. South-facing windows, also known as solar windows, are designed to maximize solar gain during winter. These windows allow sunlight to penetrate the building, warming interior spaces. To prevent overheating, overhangs or shading devices block the higher-angle summer sun while still allowing the lower-angle winter sun to enter. This careful balance ensures that passive solar design achieves optimal performance throughout the year, responding to the dynamic patterns of the sun.

Natural ventilation is another integral component of passive solar design, promoting airflow to cool and refresh indoor spaces. Operable windows strategically placed to capture prevailing winds facilitate cross-ventilation, expelling hot air and promoting passive cooling. Thermal buoyancy, driven by temperature differences between indoor and outdoor air, can be harnessed through stack ventilation. This involves the installation of high and low openings to encourage the movement of warm air upward and its replacement with cooler outdoor air. By incorporating natural ventilation strategies, passive solar buildings enhance indoor air quality and reduce reliance on mechanical cooling systems.

Day lighting, using natural light to illuminate indoor spaces, is an aesthetic and energy-efficient aspect of passive solar design. Well-designed buildings prioritize the integration of large windows, skylights, and light shelves to optimize daylight penetration. This reduces the need for artificial lighting during the day and contributes to a healthier and more visually comfortable indoor environment. Daylighting strategies consider factors such as building orientation, window size and placement, and the use of reflective surfaces to distribute natural light evenly throughout the interior.

Passive solar design extends beyond individual buildings to consider the overall layout and arrangement of structures in a development or community. Site planning and landscaping play crucial roles in optimizing solar exposure and shading. Trees, vegetation, and other natural elements can be strategically positioned to provide shade in the summer, reducing cooling demands while allowing sunlight to reach buildings in the winter. This holistic approach to site planning enhances the overall sustainability of development, creating a cohesive and energy-efficient built environment.

Passive solar principles are adaptable to various building types, from residential homes to commercial and institutional structures. The design considerations may vary based on each building type's specific requirements and functions, but the underlying principles remain consistent. In residential architecture, passive solar design can be seamlessly integrated into the layout of single-family homes or multi-unit developments. Large south-facing windows, thermal mass incorporated into flooring or walls, and effective insulation contribute to energy-efficient and comfortable living spaces. With their unique occupancy patterns and internal loads, commercial buildings benefit from passive solar design through optimized glazing, daylighting strategies, and energy-efficient HVAC systems.

In institutional architecture, such as schools or healthcare facilities, passive solar design principles create comfortable and environmentally responsive spaces. Incorporating atriums, courtyards, or glazed connectors enhances natural light penetration, creating uplifting and healing environments. Educational institutions, in particular, can leverage passive solar design to integrate sustainability principles into the built environment, fostering awareness and environmental stewardship among students and faculty.

Despite its numerous advantages, passive solar design has challenges. One of the primary challenges is the need for careful and site-specific planning. The effectiveness of passive solar design relies on factors such as climate, topography, and local solar patterns. While the principles are universally applicable, their implementation requires a nuanced understanding of the specific conditions of each location. Additionally, the balance between maximizing solar gain in the winter and minimizing overheating in the summer requires a delicate calibration of design elements and technologies.

Another challenge lies in retrofitting existing buildings to incorporate passive solar design features. While it may be more straightforward to integrate these principles into new construction, retrofitting poses logistical and structural challenges. However, advancements in building technologies, such as the development of high-performance glazing and energy-efficient retrofit solutions, are gradually making it more feasible to upgrade existing structures with passive solar features.

In conclusion, passive solar design principles represent a harmonious fusion of architectural innovation, environmental consciousness, and energy efficiency. Rooted in ancient building practices and reimagined for the modern era, passive solar design exemplifies a holistic approach to sustainable architecture. By harnessing the sun's natural energy for heating, cooling, and lighting, passive solar buildings demonstrate the potential for human habitats to exist in harmony with their natural surroundings. As the imperative for sustainable development and climate resilience grows, passive solar design stands as a beacon of inspiration, demonstrating that buildings can be functional and beautiful while treading lightly on the planet.

Insulation and heating solutions

Insulation and heating solutions play pivotal roles in the pursuit of energy efficiency, comfort, and environmental sustainability in the built environment. These elements contribute significantly to a building's thermal performance, affecting its energy consumption and the ease of its occupants. As societies increasingly prioritize sustainable living and address the challenges of climate change, integrating effective insulation and heating solutions becomes imperative. This section explores the fundamental principles, materials, and technologies associated with insulation and heating, examining how thoughtful design and innovative solutions contribute to energy-efficient and environmentally conscious buildings.

Insulation serves as a critical component in the effort to regulate a building's thermal environment. The primary function of insulation is to resist heat flow, preventing heat transfer between the interior and exterior of a building. By minimizing heat exchange, insulation enhances a building's energy efficiency, reducing the need for artificial heating or cooling. Various insulation materials are available, each with unique properties and applications. Fiberglass insulation, composed of fine glass fibers, is a common choice due to its affordability and effectiveness in preventing heat transfer. Another widely used material is cellulose insulation, made from recycled paper or plant fibers, providing an environmentally friendly option with good thermal performance.

Spray foam insulation is a versatile and efficient solution for thermal and air barrier applications. This expanding foam conforms to irregular shapes and penetrations, creating a seamless, air-tight barrier. This reduces heat transfer and prevents drafts and air leakage, enhancing overall energy efficiency. Additionally, natural insulation materials, such as wool or cotton, offer renewable and biodegradable alternatives, catering to those seeking environmentally conscious building practices.

Effective insulation involves selecting the suitable material and ensuring proper installation. Gaps, voids, or compression in insulation can compromise its performance. Hence, meticulous attention to installation details is essential to maximize the effectiveness of insulation in maintaining a comfortable indoor environment. Advances in building science have led to improved insulation techniques, such as continuous insulation systems that minimize thermal bridging and enhance overall energy performance.

Heating solutions are intrinsically linked to insulation, as a well-insulated building requires less energy to maintain a comfortable temperature. However, the choice of heating systems significantly influences a building's energy consumption and environmental impact. Conventional heating systems, such as furnaces or electric heaters, have been common choices, but their reliance on fossil fuels or grid electricity contributes to greenhouse gas emissions. There is a growing emphasis on sustainable heating solutions that harness renewable energy sources and promote energy efficiency.

One prominent example is radiant heating, which delivers heat directly to the occupants or surfaces within a space. This method is often achieved through underfloor heating systems or radiant panels on walls or ceilings. Radiant heating offers several advantages, including even distribution of heat, reduced air stratification, and the potential for lower operating temperatures compared to traditional forced-air systems. Additionally, incorporating renewable energy sources, such as solar thermal systems, enhances the sustainability of radiant heating solutions.

Geothermal heating systems utilize the consistent temperature of the ground to provide heating and cooling for buildings. Ground-source heat pumps transfer heat between the ground and the building, using the earth's stable temperature to achieve energy efficiency. While geothermal systems may require an initial investment, they offer long-term benefits in terms of reduced operating costs and lower environmental impact. Air-source heat pumps, another popular option, extract heat from the outdoor air and transfer it indoors for heating. Advances in heat pump technology, including variable-speed compressors and improved efficiency, make these systems viable alternatives for sustainable heating.

Solar heating technologies harness the sun's energy to provide warmth for buildings. Passive solar heating utilizes design strategies, such as south-facing windows and thermal mass, to naturally capture and store solar heat. On the other hand, active solar heating systems incorporate solar collectors to absorb sunlight and transfer the heat to a fluid, which is then used for space heating. Integrating solar heating into building design aligns with passive solar principles, creating a symbiotic relationship between architecture and renewable energy.

Biomass heating systems utilize organic materials, such as wood pellets, agricultural residues, or dedicated energy crops, as fuel sources for heating. Biomass is considered a renewable energy source as long as the consumption rate does not exceed the replenishment rate. Modern biomass heating technologies, including high-efficiency stoves and pellet boilers, offer cleaner combustion and reduced emissions compared to traditional wood-burning stoves. Biomass heating can be a sustainable option, especially in regions with abundant biomass resources.

District heating systems represent a community-level approach to heating, where a centralized plant produces heat and distributes it to multiple buildings through a network of pipes. These systems often utilize combined heat and power (CHP) technologies, maximizing the efficiency of energy production. District heating can incorporate various heat sources from industrial processes, including biomass, geothermal, or waste heat. District heating reduces individual building emissions and promotes energy sharing within communities by centralizing heating production.

Intelligent heating controls and building automation systems contribute to the overall efficiency of heating solutions. Smart thermostats, occupancy sensors, and adaptive control algorithms optimize heating based on real-time conditions and user preferences. These systems enable precise temperature control, minimize energy waste, and enhance the responsiveness of heating systems to changing environmental factors. Integrating innovative technologies aligns with the broader trend of creating intelligent, connected buildings that adapt to user needs and environmental conditions.

Challenges persist in the widespread adoption of sustainable insulation and heating solutions. Economic considerations, initial investment costs, and the availability of renewable energy sources may influence decision-making in both residential and commercial construction. Overcoming these challenges requires a holistic approach considering the life cycle costs, long-term benefits, and environmental impacts of insulation and heating choices. Government incentives, regulations promoting energy-efficient building practices, and increased public awareness of the benefits of sustainable construction contribute to creating a supportive environment for adopting these solutions.

In conclusion, insulation and heating solutions are integral to energy-efficient and environmentally conscious building design. Thoughtful insulation choices and sustainable heating technologies contribute to reduced energy consumption, lower greenhouse gas emissions, and enhanced occupant comfort. As the global community continues to grapple with the challenges of climate change, the importance of adopting insulation and heating solutions that align with sustainability principles cannot be overstated. By integrating these solutions into building design and construction practices, societies can contribute to a more resilient and sustainable built environment for current and future generations.

Building a sustainable off-grid home

Building sustainable off-grid homes has gained prominence as the world grapples with the urgent need to address climate change and minimize environmental impact. An off-grid home operates independently of traditional utility services, relying on self-sufficient systems to generate energy, manage water resources, and maintain a comfortable living environment. This approach aligns with the principles of environmental stewardship and offers individuals the opportunity to embrace a lifestyle characterized by autonomy and reduced reliance on external infrastructure. This section explores the fundamental principles and key considerations involved in building a sustainable off-grid home, emphasizing the integration of renewable energy sources, efficient design strategies, and responsible resource management.

At the core of a sustainable off-grid home is the judicious use of renewable energy sources to meet the household's energy demands. Solar power is a primary and widely adopted renewable energy option for off-grid living. Photovoltaic (PV) panels, strategically positioned on the roof or ground-mounted arrays, convert sunlight into electricity. This clean and abundant energy source not only powers essential appliances and lighting but can also be stored in batteries for use during periods of low sunlight. Advanced solar technology and decreasing costs have made solar power an accessible and efficient solution for off-grid homeowners seeking energy independence.

Wind energy presents another viable option for off-grid electricity generation. Small-scale wind turbines in areas with consistent wind patterns can contribute to a home's energy needs. Wind power complements solar energy by providing electricity during periods of low sunlight or at night. Integrating solar and wind systems in a hybrid approach enhances the reliability and resilience of off-grid power systems. However, considering local wind conditions, potential noise issues, and visual impact is essential when incorporating wind energy into off-grid home designs.

Micro-hydro systems harness the energy from flowing water to generate electricity. Micro-hydro systems offer a consistent and reliable energy supply for properties with access to a water source, such as a stream or river. These systems use the kinetic energy of moving water to turn a turbine, generating electricity that can be stored or used directly. Micro-hydro systems are particularly advantageous in off-grid settings where abundant water resources provide a continuous and sustainable power source.

Efficient design strategies are crucial in maximizing the performance of renewable energy systems in off-grid homes. Passive solar design principles, which optimize a building's orientation, window placement, and thermal mass to capture and retain solar heat, contribute to natural heating and cooling. This reduces the energy demand for climate control, enhancing the overall efficiency of off-grid living. Thoughtful placement of windows and insulation materials, coupled with attention to natural ventilation, helps regulate indoor temperatures and minimizes the need for mechanical heating or cooling systems.

Energy-efficient appliances and lighting further contribute to the sustainability of off-grid homes. Choosing appliances with high energy efficiency ratings, utilizing LED or other energy-saving lighting technologies, and adopting smart home controls enhance overall energy performance. Off-grid homeowners often employ energy monitoring systems to track and optimize energy usage, allowing for informed decision-making and resource management.

Water management is a critical consideration in sustainable off-grid home construction. Rainwater harvesting systems, consisting of roof collection surfaces, gutters, and storage tanks, capture and store rainwater for domestic use. This decentralized approach reduces reliance on external water sources and minimizes the environmental impact of centralized water supply systems. Greywater recycling systems, which treat and reuse water from activities like bathing and laundry, contribute to responsible water usage and conservation. These systems can be integrated into landscaping or used for non-potable purposes, reducing the strain on water resources.

Composting toilets represent an eco-friendly alternative to traditional flush toilets, especially in off-grid settings. These systems convert human waste into nutrient-rich compost, eliminating the need for water-intensive sewage systems. Composting toilets contribute to sustainable waste management, promoting resource recovery and reducing the environmental footprint associated with conventional sewage treatment.

Material selection plays a pivotal role in sustainable off-grid home construction. Locally sourced, recycled, or reclaimed materials contribute to a lower carbon footprint and support regional economies. Building materials with low embodied energy, such as bamboo, straw bales, or rammed earth, align with sustainable construction practices. Timber from responsibly managed forests, certified by organizations like the Forest Stewardship

Council (FSC), ensures that wood products come from environmentally and socially responsible sources.

Incorporating passive design features further enhances the sustainability of off-grid homes. Green roofs, covered with vegetation, provide insulation, reduce stormwater runoff, and contribute to biodiversity. The thermal mass of earthen materials or other sustainable building materials helps regulate indoor temperatures, reducing the need for active heating or cooling. Proper insulation and well-sealed windows and doors prevent heat loss in colder climates and maintain a comfortable indoor environment throughout the year.

Integrating permaculture principles into the design of off-grid homes promotes self-sufficiency and ecological harmony. Permaculture, a design philosophy centered on sustainable and regenerative practices, encompasses carefully planning food production, water management, and waste recycling within the home and its surroundings. Integrating edible landscapes, rain gardens, and agroforestry practices into off-grid properties enhances resilience, biodiversity, and the overall sustainability of the living environment.

Waste reduction and recycling practices are fundamental components of a sustainable off-grid lifestyle. Minimizing single-use items, adopting a "zero-waste" mindset, and implementing effective recycling systems contribute to responsible resource management. Composting organic waste, reusing materials, and repurposing items for various functions exemplify the ethos of sustainable living and resource conservation in off-grid contexts.

Community engagement and knowledge-sharing play essential roles in developing sustainable off-grid communities. Establishing networks for information exchange, collaborative projects, and shared resources fosters a sense of community and resilience.

Off-grid communities often organize events, workshops, and skill-sharing sessions to empower residents with the knowledge and skills needed to maintain and improve their off-grid lifestyles.

Challenges exist in pursuing sustainable off-grid living, and solutions often require a balance between technological innovation, lifestyle choices, and responsible resource management. Initial costs associated with renewable energy systems, water harvesting infrastructure, and sustainable building materials may present barriers to entry. However, the long-term economic, environmental, and lifestyle benefits often outweigh these upfront investments. Regulatory frameworks, supportive policies, and financial incentives can play pivotal roles in facilitating the adoption of sustainable off-grid practices, encouraging innovation, and broadening access to these alternative living solutions.

In conclusion, building a sustainable off-grid home is a multifaceted endeavor encompassing renewable energy systems, efficient design strategies, responsible water management, and a commitment to sustainable living principles. The integration of these elements reduces the environmental impact of human habitation and fosters a lifestyle characterized by self-sufficiency, resilience, and a harmonious relationship with the natural world. As the global community grapples with the challenges of climate change and resource depletion, the principles and practices of sustainable off-grid living offer a glimpse into a future where individuals and communities thrive in balance with the planet.

CHAPTER V

Food Production

Organic gardening and permaculture

In the quest for sustainable and regenerative agricultural practices, organic gardening and permaculture have emerged as influential paradigms, promoting ecological harmony, biodiversity, and resilient food systems. Organic gardening involves cultivating plants without synthetic pesticides, herbicides, or fertilizers, emphasizing natural and holistic approaches to soil health and plant nutrition. Permaculture, short for "permanent agriculture" or "permanent culture," is a design philosophy that integrates diverse agricultural, ecological, and social principles to create harmonious and autonomous systems. Together, these approaches offer a holistic framework for cultivating food that nurtures the environment, supports local ecosystems, and provides nutritious produce. This section delves into organic gardening and permaculture principles and practices, exploring how they contribute to sustainable agriculture and the broader goal of creating resilient and regenerative food systems.

Organic gardening, at its core, rejects synthetic chemicals in favor of natural and sustainable methods. The emphasis is on building and maintaining healthy soil, recognizing it as the foundation for robust plant growth. Organic gardeners prioritize using compost, cover crops, and organic mulches to enhance soil fertility, structure, and water retention. By fostering a thriving soil ecosystem, organic gardens create a balanced, nutrient-rich environment supporting plant health and resilience.

Crop rotation is fundamental in organic gardening, helping prevent soil-borne diseases and pests. This method involves alternating the types of crops planted in specific areas over time, disrupting the life cycles of potential pathogens and insects. Companion planting is another strategy employed to enhance biodiversity and deter pests. Certain plant combinations have symbiotic relationships, where one species may repel pests that affect the other or enhance nutrient uptake through root interactions. These synergies contribute to the garden's overall health and reduce reliance on external inputs.

The rejection of synthetic pesticides and herbicides in organic gardening aligns with a commitment to environmental health. Instead, organic gardeners employ biological pest control methods, such as introducing beneficial insects that prey on harmful pests or using natural predators to maintain ecological balance. This approach minimizes harm to non-target species and avoids the negative environmental impact associated with chemical interventions. Integrated pest management (IPM) principles guide organic gardeners in adopting a proactive and holistic approach to pest control, considering the entire ecosystem and its dynamics. Permaculture extends the principles of organic gardening to encompass broader design concepts, recognizing the interconnectedness of agriculture, ecology, and human systems. Developed by Bill Mollison and David Holmgren in the 1970s, permaculture seeks to create sustainable and resilient human habitats by mimicking natural ecosystems and utilizing regenerative design principles. The three core ethics of permaculture—earth care, people care, and fair share—underscore a commitment to environmental stewardship, community well-being, and equitable resource distribution.

Permaculture design principles emphasize the thoughtful placement of elements within a system to maximize their beneficial relationships. Zones and sectors are used to organize and design spaces based on the frequency of human interaction and the flow of energy and resources. For example, high-maintenance and frequently visited areas are placed closer to the home (Zone 1), while less intensively managed areas, such as orchards or food forests, may be located in more distant zones. This zoning approach allows for efficient management and utilization of resources.

Agroforestry, a key component of permaculture, integrates trees and perennial plants into agricultural landscapes. This approach enhances biodiversity and provides multiple yields, such as fruits, nuts, timber, and habitat for beneficial wildlife. Forest gardening, a specific form of agroforestry, mimics the structure and function of natural forests, utilizing layers of vegetation to create productive and resilient ecosystems. Permaculture principles guide the design of polyculture systems, where diverse plant species are interplanted to enhance soil fertility, minimize pest pressures, and create resilient food systems.

Water management is critical in permaculture design, emphasizing capturing, storing, and efficiently utilizing water resources. Rainwater harvesting systems, swales, and contour planting help slow down water runoff, allowing it to infiltrate and replenish the soil. Earthworks, such as ponds or berms, serve multiple functions, including water storage, flood control, and habitat creation. By integrating water management strategies, permaculture designs create landscapes that are resilient to climatic variations and contribute to the conservation of water resources.

Permaculture also places a strong emphasis on closed-loop systems and waste reduction. The concept of "stacking functions" encourages the design of elements that serve multiple purposes. For example, a chicken coop may provide eggs and contribute to pest control through foraging and help build soil fertility by incorporating their manure. Additionally, permaculture promotes the use of recycled or repurposed materials, reducing the environmental impact of construction and encouraging a culture of resourcefulness.

Social permaculture extends the principles of permaculture beyond the physical landscape to consider human interactions and community dynamics. This includes fostering community resilience, participatory decision-making, and the development of local economies. Community gardens, cooperatives, and shared resources exemplify social permaculture principles, creating spaces where individuals collaborate, share knowledge, and collectively care for the environment.

While organic gardening and permaculture are committed to sustainable and regenerative practices, they also face challenges and critiques. Critics argue that the yields from organic and permaculture systems may be lower than those achieved through conventional agriculture, potentially limiting their scalability to feed growing populations. Additionally, the upfront investment and knowledge required for permaculture design may pose barriers to widespread adoption, particularly in regions where traditional agricultural practices are deeply ingrained.

In conclusion, organic gardening and permaculture offer holistic and sustainable approaches to agriculture, emphasizing the interconnectedness of natural systems and human well-being. These practices prioritize environmental health, biodiversity, and resource efficiency, creating resilient food systems that adapt to changing environmental conditions. As global challenges such as climate change, resource depletion, and food insecurity become more pronounced, organic gardening and permaculture principles and practices provide valuable insights into creating regenerative and sustainable solutions for the future. By embracing these approaches, individuals, communities, and societies can create a more harmonious and resilient relationship between humans and the natural world.

Hydroponics and aquaponics

In pursuing sustainable and efficient agricultural practices, hydroponics and aquaponics have emerged as innovative and resource-efficient systems that redefine traditional approaches to farming. These soil-less cultivation methods leverage technology to optimize water usage, nutrient delivery, and space utilization, offering solutions to challenges such as water scarcity and limited arable land. Hydroponics involves growing plants in a nutrient-rich water solution, while aquaponics combines hydroponics with aquaculture, creating a symbiotic relationship between plants and aquatic animals. This section explores the principles, benefits, and challenges associated with hydroponics and aquaponics, showcasing their potential to revolutionize modern agriculture and contribute to global food security.

Hydroponics, derived from the Greek words "hydro" (water) and "ponos" (labor), represents a cultivation method that relies on a nutrient solution rather than soil to deliver essential elements to plants. In hydroponic systems, plants are grown in inert growing mediums such as perlite, coconut coir, or Rockwool, allowing the roots to come into direct contact with the nutrient solution. This method enables precise control over nutrient levels, pH, and environmental conditions, facilitating optimal plant growth. Hydroponic systems come in various forms, including nutrient film technique (NFT), deep water culture (DWC), aeroponics, and drip systems, each offering unique advantages and applications.

One of the primary benefits of hydroponics is the ability to maximize resource efficiency. Unlike traditional soil- based agriculture, where plants expend energy searching for nutrients, hydroponic systems deliver nutrients directly to the root zone, minimizing waste and increasing nutrient uptake. The controlled environment allows for year-round cultivation, independent of seasonal variations, resulting in higher crop yields and faster growth rates. Additionally, hydroponic systems use significantly less water than traditional farming, making them well-suited for regions facing water scarcity challenges.

The precise control over growing conditions in

hydroponics extends beyond nutrient management. pH levels can be adjusted to suit the specific needs of different crops, ensuring optimal nutrient absorption. Temperature, humidity, and light intensity can also be fine-tuned, creating an ideal environment for plant growth. This level of control reduces the reliance on pesticides and herbicides, making a more environmentally friendly and sustainable form of agriculture.

Hydroponics has proven particularly advantageous in

urban agriculture, where limited space and soil quality pose challenges. Vertical farming, a form of hydroponics, utilizes vertical space to stack layers of crops, maximizing production within a confined area. This approach has gained traction in densely populated urban areas, contributing to local food production and reducing the environmental footprint of transporting food long distances.

While hydroponics excels in resource efficiency and

controlled environments, aquaponics takes sustainability further by integrating fish farming with hydroponic plant cultivation. Aquaponics creates a closed-loop system where the waste produced by aquatic animals, typically fish, serves as a nutrient source for plants. The plants filter and purify the water before recirculating it to the fish tanks. This mutualistic relationship mimics natural ecosystems, providing a holistic and sustainable approach to food production.

The cornerstone of aquaponics is the nitrogen cycle,

where fish excrete ammonia as waste. Beneficial bacteria in the system convert ammonia into nitrites and then into nitrates, which serve as the primary nutrient source for plants. As the plants absorb these nutrients, they help purify the water, creating a symbiotic and self-sustaining ecosystem. Common fish species used in aquaponics include tilapia, trout, and catfish, chosen for their adaptability to aquaponic environments.

The integration of aquaponics offers several key advantages. Firstly, it significantly reduces the need for external fertilizers, as the fish waste provides a natural and continuous nutrient source. Secondly, the system conserves water, with the same water being recirculated between the fish tanks and plant beds. This closed-loop approach minimizes water usage and creates a more environmentally sustainable method of agriculture, particularly relevant in regions facing water scarcity. Thirdly, aquaponics allows for cultivating fish and plants in a single system, diversifying the outputs and providing a balanced and nutritious yield.

Beyond its environmental benefits, aquaponics provides a unique model for small-scale and backyard farming. The system can be scaled to suit various needs, from small home setups to more extensive commercial operations. Aquaponic kits and DIY systems have become increasingly accessible, empowering individuals to engage in sustainable food production at a personal level. The educational value of aquaponics is also notable, as it provides a hands-on and interdisciplinary approach to learning about biology, chemistry, and environmental science.

While hydroponics and aquaponics offer numerous advantages, challenges require careful consideration and management. The initial setup costs for hydroponic and aquaponic systems can be higher than traditional farming methods, posing a barrier for some farmers. However, as technology advances and adoption increases, economies of scale and improved efficiency may contribute to the cost-effectiveness of these systems. Additionally, hydroponics and aquaponics require specific technical knowledge and expertise, necessitating training and education for successful implementation.

Maintaining water quality is crucial in aquaponics, as any imbalance can negatively impact fish and plant health. Monitoring parameters such as pH, ammonia levels, and nutrient concentrations is essential for the system's stability. Disease management in fish populations and pest control in hydroponic systems also require attention to prevent outbreaks that could harm the entire ecosystem. Integrating backup systems and redundancy measures is critical to mitigate the risks associated with system failures.

In conclusion, hydroponics and aquaponics stand as innovative and sustainable alternatives to traditional farming methods, addressing resource efficiency, water scarcity, and urban agriculture challenges. These soil-less cultivation techniques provide a blueprint for the future of agriculture, offering precise control over growing conditions, minimizing environmental impact, and promoting closed-loop systems. As the global population continues to grow and the demand for food rises, embracing technologies like hydroponics and aquaponics can contribute to building resilient and sustainable food systems. Through ongoing research, education, and technological advancements, these systems have the potential to revolutionize the way we approach food production, making it more efficient, environmentally friendly, and accessible to diverse communities worldwide.

Livestock for self-sufficiency

Integrating livestock into a self-sufficient lifestyle represents a time-honored and practical approach to meeting diverse needs, ranging from food production to resource utilization. Livestock, including cattle, sheep, goats, chickens, pigs, and more, serve as valuable assets for individuals and communities striving to achieve self-sufficiency. This section explores the multifaceted roles of livestock in supporting self-sufficiency, encompassing aspects of food security, resource utilization, and sustainable agriculture.

One of the primary benefits of raising livestock for self-sufficiency lies in producing high-quality, protein-rich food. Livestock provide a renewable source of meat, eggs, and dairy products, contributing essential nutrients to the diet. Meat from animals like cattle, goats, and chickens is a valuable protein source that complements plant-based foods, creating a well-rounded and nutritionally rich diet. Eggs from poultry and milk from dairy animals further enhance the variety and nutritional content of self- produced food, reducing reliance on external sources.

In addition to food production, livestock contributes to resource utilization through their ability to convert non-human edible feedstuffs into valuable products. Ruminants, such as cattle and sheep, can convert fibrous plant materials, including grasses and crop residues, into meat and milk through enteric fermentation. This ecological adaptation allows livestock to thrive on land that may not be suitable for crop cultivation, thereby expanding the productive capacity of a self-sufficient homestead.

Livestock also plays a pivotal role in sustainable agriculture by contributing to nutrient cycling. Animal manure is a natural fertilizer, replenishing soil with essential nutrients such as nitrogen, phosphorus, and potassium. This closed-loop system supports healthy soil structure and fertility, enhancing the productivity of cultivated crops. Integrating livestock with crop farming in a well-managed rotational system exemplifies the principles of agroecology, where the synergy between plants and animals creates a balanced and regenerative agricultural ecosystem.

Chickens, in particular, showcase the symbiotic relationship between livestock and crop production through their role in pest control. Chickens are natural foragers that consume insects, weeds, and kitchen scraps, reducing the need for chemical pesticides and herbicides. Integrating chickens into a permaculture or agroecological system helps control pests, enhances soil health, and contributes to a more resilient and self-sufficient farming model.

Livestock provide valuable byproducts beyond meat and dairy, contributing to various aspects of self-sufficiency. Wool from sheep, for example, offers a renewable and natural fiber source for clothing and textiles. Goats can provide both meat and fiber through their cashmere or mohair production. Pigs contribute to self-sufficiency by efficiently converting kitchen and garden waste into compost and edible products. Recognizing and maximizing livestock's diverse outputs enhances the sustainability and resilience of a self-sufficient lifestyle.

The concept of "stacking functions" within a self-sufficient homestead involves integrating livestock into multiple aspects of the ecosystem. For example, using chickens to scratch and turn compost helps manage organic waste and provides an opportunity for the birds to forage and contribute to pest control. Similarly, rotational grazing of cattle on pasture not only produces meat and milk but also promotes healthy grass growth and soil fertility. This integrated approach exemplifies the interconnectedness of livestock management and sustainable land use.

Animal husbandry practices in a self-sufficient context extend beyond conventional livestock to include smaller and more manageable species. Rabbits, for instance, are prolific breeders and efficient converters of plant material into protein. Their small size and low space requirements make them well-suited for backyard or small-scale self-sufficient setups. Though not traditional livestock, bees play a vital role in pollination and honey production, contributing to crop yields and providing a sweetener for self-produced foods.

Integrating livestock into a self-sufficient lifestyle requires careful planning and consideration of animal welfare, environmental impact, and the specific needs of the homestead. Adequate space, appropriate shelter, and proper nutrition are essential for responsible livestock management. Pasture rotation, rotational grazing, and agroecological design principles help optimize the use of available land while preventing overgrazing and soil degradation. Providing animals with access to diverse forage, clean water, and suitable living conditions ensures their well-being and enhances their ability to contribute to the self-sufficiency of the homestead.

Challenges exist in balancing the benefits of livestock integration with the responsibilities and ethical considerations of animal care. Issues such as overgrazing, waste management, and the potential environmental impact of concentrated animal feeding operations (CAFOs) highlight the importance of adopting sustainable and humane husbandry practices. Individuals can mitigate the adverse effects of conventional industrial livestock farming by embracing regenerative agriculture principles, such as holistic management and agroecology.

In conclusion, livestock play a central and multifaceted role in achieving self-sufficiency, offering valuable contributions to food production, resource utilization, and sustainable agriculture. Integrating livestock into a self-sufficient lifestyle enhances the resilience of homesteads, diversifies food sources, and supports nutrient cycling in agroecological systems. While challenges exist in managing livestock farming's environmental impact and ethical considerations, adopting responsible and regenerative practices contributes to a more balanced and sustainable approach to animal agriculture. As individuals and communities explore self-sufficiency as a pathway to a more resilient and sustainable future, the thoughtful integration of livestock becomes a cornerstone in creating holistic and harmonious ecosystems that support human needs while respecting the broader environment.

Preserving and storing food off-grid

In the pursuit of self-sufficiency and resilience, preserving and storing food off-grid emerges as a fundamental skill that empowers individuals and communities to manage their resources efficiently and withstand uncertainties. Off-grid living, whether driven by a desire for sustainability, preparedness, or a connection to nature, demands a comprehensive approach to food preservation to ensure a year-round supply of nourishment. This section delves into various methods and strategies for preserving and storing food off-grid, exploring traditional techniques alongside modern innovations that harmonize with the principles of sustainable living.

Canning, a time-honored method of food preservation, involves sealing food in jars and heat-processing them to destroy microorganisms that cause spoilage. This method is particularly effective for fruits, vegetables, jams, and sauces. The process creates a vacuum seal that prevents the entry of air and inhibits the growth of bacteria. Similarly, fermentation harnesses the power of beneficial microorganisms to preserve food. Vegetables, such as sauerkraut and kimchi, and dairy products, like yogurt and kefir, undergo a natural fermentation process that not only extends their shelf life but also enhances their nutritional value. Using brine or vinegar, pickling adds a tangy flavor to preserved foods, including cucumbers, beets, and various fruits.

Drying and dehydrating food is a timeless method that removes moisture, inhibiting the growth of bacteria and molds. Sun drying, a traditional approach, involves exposing food to the sun's warmth for extended periods. Modern dehydrators, powered by electricity or alternative energy sources, offer a controlled environment for drying a wide range of foods, including fruits, vegetables, herbs, and meats. Dehydrated foods are lightweight, space- efficient, and retain much of their nutritional value, making them ideal for off-grid storage. The process concentrates flavors, providing a convenient and flavorful addition to off-grid meals.

Root cellars represent a traditional and low-tech solution for preserving fruits and vegetables without the need for energy-intensive methods. These underground storage spaces leverage the natural coolness of the earth to maintain a stable temperature and humidity level, creating an environment conducive to long-term food storage. Vegetables like potatoes, carrots, and apples can be stored in root cellars, offering a simple yet effective way to extend their shelf life. In addition to root cellars, cold storage methods utilize cool and dark spaces, such as basements or shaded outdoor areas, to slow produce's natural ripening and decay.

Smoking and salting are age-old techniques for preserving meats, imparting distinct flavors while inhibiting microbial growth. Smoking exposes meats to the natural preservatives in wood smoke, enhancing their longevity. Fish, poultry, and certain cuts of meat are commonly preserved through smoking. Salting involves coating meats with salt, which draws out moisture and creates an inhospitable environment for bacteria. Salted meats, like bacon and salted fish, have been staples in traditional diets for centuries. These methods contribute to food preservation and add unique and savory characteristics to off-grid culinary endeavors.

In the context of off-grid living, modern innovations have introduced sustainable alternatives that align with the principles of self-sufficiency. Solar dehydrators harness the power of the sun to dry and preserve foods, offering an energy-efficient solution for off-grid environments. Electric-free refrigeration systems, such as zeer pots or pot-in-pot coolers, use evaporation to maintain lower temperatures, providing a simple and accessible means of cooling perishables without relying on electricity. These innovations highlight the adaptability of off-grid living to incorporate sustainable technologies that enhance food preservation capabilities.

Hermetic sealing, often used with vacuum packaging, involves removing air from storage containers to create an airtight seal. This method prevents the oxidation and deterioration of food, maintaining freshness and nutritional quality. Vacuum packaging, facilitated by manual or electric vacuum sealers, extends the shelf life of various foods, from grains and legumes to meats and vegetables. Hermetic sealing and vacuum packaging reduce the risk of spoilage by minimizing exposure to oxygen, making them valuable techniques for off-grid storage where traditional refrigeration may be limited.

Successful off-grid food preservation requires careful consideration of factors such as climate, available resources, and the specific needs of the homestead. Solar dehydrators may be particularly effective in regions with abundant sunlight, while colder climates might benefit from root cellars or brutal storage methods. The selection of preservation methods should align with the types of foods grown or acquired, ensuring a diverse and balanced off-grid pantry.

Efficient storage solutions also play a crucial role in off-grid food preservation. Mason jars, a versatile and reusable option, are commonly used for canning and preserving liquids. Mylar bags, vacuum-sealed pouches, and airtight containers provide options for storing dehydrated or freeze-dried foods, preventing exposure to moisture and air. Proper labeling and organization contribute to an adequate food storage system, allowing individuals to manage inventory and rotate items to maintain freshness.

While off-grid food preservation offers numerous benefits, including increased self-sufficiency, reduced reliance on external food sources, and the ability to enjoy homegrown or locally sourced produce year-round, challenges exist. Limited access to electricity may restrict the use of electric dehydrators or freezers, emphasizing the need for alternative methods. Additionally, the learning curve associated with mastering various preservation techniques requires dedication and a willingness to experiment.

Despite these challenges, the rewards of off-grid food preservation are significant. Preserving and storing food off-grid provides a sense of autonomy and security and fosters a deeper connection to the food production process. It encourages a mindful approach to resource management, minimizing waste and maximizing the utility of homegrown or locally sourced foods. Furthermore, off-grid food preservation aligns with sustainable living principles, promoting resilience in the face of environmental uncertainties and reducing the ecological footprint associated with conventional food supply chains.

In conclusion, preserving and storing food off-grid is a multifaceted and essential component of self-sufficiency. The diverse methods available, from traditional techniques like canning and fermentation to modern innovations such as solar dehydrators and hermetic sealing, offer various options to suit multiple off-grid contexts. By embracing these preservation methods, individuals and communities can extend the shelf life of their harvests, cultivate a deeper connection to the seasons, reduce reliance on external food sources, and foster a sustainable and resilient approach to off-grid living. As the world grapples with the challenges of a changing climate and uncertainties in global food systems, the art of off-grid food preservation stands as a beacon of self-reliance and a testament to the ingenuity of those living in harmony with the land.

CHAPTER VI

Waste Management

Composting and recycling

In the realm of off-grid living, where the pursuit of self-sufficiency intersects with a commitment to environmental stewardship, composting, and recycling emerge as foundational practices. These approaches epitomize sustainability principles by transforming organic and recyclable materials into valuable resources, mitigating waste, and contributing to regenerative cycles. This section explores the symbiotic relationship between off-grid living, composting, and recycling, shedding light on how these practices intertwine to cultivate a more sustainable and resilient way of life.

Composting is a cornerstone of sustainable waste management in off-grid settings, converting kitchen scraps, yard waste, and organic matter into nutrient-rich soil amendments. Composting mimics natural decomposition processes at its core, harnessing the collective power of microorganisms, bacteria, and fungi to break down organic materials. Off-grid dwellers often utilize compost bins, tumblers, or designated compost heaps to facilitate decomposition.

In the off-grid context, composting is a holistic endeavor that aligns with permaculture and regenerative agriculture principles. Kitchen scraps such as fruit and vegetable peels, coffee grounds, and eggshells contribute to compost fertility. Yard waste, including leaves, grass clippings, and prunings, adds bulk and aeration. Composting also accommodates organic materials like cardboard and newspaper, introducing carbon-rich components to balance the nitrogen-rich kitchen and yard waste. The resulting compost becomes a valuable soil conditioner, enhancing fertility, water retention, and microbial activity in gardens and agricultural plots.

Composting in off-grid environments fosters a cyclical relationship with the land. Food scraps return to the soil, enriching with nutrients that nurture future crops. This closed-loop system minimizes the need for external inputs, reduces reliance on synthetic fertilizers, and exemplifies a sustainable approach to soil management. In regions where access to commercial fertilizers may be limited, composting becomes a vital practice for maintaining soil health and promoting successful harvests.

Recycling complements composting in the off-grid pursuit of sustainability, addressing the challenge of managing non-organic waste materials. While recycling systems vary across regions, the fundamental concept remains consistent: transforming used materials into new products. In off-grid settings, where infrastructure for waste collection and disposal may be limited, residents often adopt creative and resourceful approaches to recycling.

Common recyclable materials include paper, cardboard, glass, metal, and certain plastics. Off-grid inhabitants often sort and store recyclables, taking advantage of local recycling facilities when available. In some cases, recycling efforts extend beyond conventional materials, incorporating repurposing and upcycling into the ethos of off-grid living. Items like glass jars, metal containers, and wooden pallets find new life as storage containers, building materials, or functional art.

Recycling in off-grid contexts embodies the principles of resourcefulness and resilience. Off-grid dwellers reduce their environmental impact and contribute to a circular economy by diverting materials from landfills and incinerators. Recycling also aligns with the ethos of minimalism and conscious consumption, encouraging individuals to consider the life cycle of products and make choices that prioritize durability and recyclability.

The synergy between composting and recycling in off-grid living lies in their collective contribution to a holistic and sustainable lifestyle. The organic waste diverted through composting enriches the soil and reduces the volume of material destined for landfills. Meanwhile, recycling addresses the non-organic waste stream, ensuring that materials such as paper, plastic, and metal are repurposed rather than discarded.

In the off-grid ethos, the integration of composting and recycling extends beyond waste management to embody a broader commitment to environmental responsibility. By cultivating nutrient-rich compost for soil health and diverting recyclables from the waste stream, off-grid practitioners actively engage in practices that mitigate ecological degradation and promote a regenerative relationship with the natural world. This conscious approach to waste aligns with the ethos of permaculture, which emphasizes the importance of designing human systems in harmony with natural ecosystems.

While composting and recycling are pillars of sustainable living, off-grid environments present unique challenges that require innovative solutions. Limited access to municipal recycling facilities may necessitate exploring creative recycling methods, such as repurposing materials on-site or finding local recyclable outlets. Composting systems must adapt to diverse climates, considering factors like moisture levels, temperature variations, and available space.

Innovations in off-grid composting and recycling include the development of composting toilets, which transform human waste into safe and nutrient-rich compost for non-edible plants. Additionally, community-based initiatives promote the sharing and repurposing items, reducing the need for new products and minimizing waste. Off-grid communities often embrace a culture of repair and reuse, extending the lifespan of goods and minimizing the environmental impact of consumerism.

Composting and recycling in off-grid settings extend beyond individual practices to become catalysts for community building and education. Off-grid communities often share knowledge and best practices in waste management, fostering a collective commitment to sustainability. Workshops and educational programs on composting and recycling empower residents with the skills and understanding needed to implement effective waste reduction strategies.

Engaging in composting and recycling also prompts a deeper reflection on consumption patterns and the ecological footprint of daily choices. Off-grid living inherently encourages mindfulness about resource use, driving individuals to consider the life cycle of products, make informed purchasing decisions, and actively participate in reducing waste generation.

In conclusion, composting and recycling are integral to off-grid living, embodying sustainability principles, resilience, and environmental responsibility. The synergy between these practices creates a harmonious waste management system that addresses organic and non-organic materials. In off-grid environments, where self-sufficiency and connection to nature are paramount, composting enriches the soil and supports regenerative agriculture, while recycling minimizes the environmental impact of waste.

The challenges presented by off-grid living, including limited infrastructure and diverse environmental conditions, inspire creative and innovative solutions. Composting toilets, community-based initiatives, and educational programs demonstrate the adaptability and resourcefulness of off-grid communities in addressing waste management challenges. As the global community grapples with environmental issues and the need for sustainable living practices, the off-grid model of composting and recycling offers valuable insights and inspiration for cultivating a more harmonious and regenerative relationship with the planet.

Sustainable waste disposal methods

In the wake of escalating environmental concerns and the pressing need for conscientious resource management, sustainable waste disposal has emerged as a critical aspect of responsible living. As societies grapple with the repercussions of excessive waste generation and the strain on natural ecosystems, exploring alternative waste disposal methods becomes imperative. This section delves into a comprehensive examination of sustainable waste disposal, unraveling a spectrum of methods that extend beyond conventional practices and usher in a paradigm shift towards environmental harmony.

At the forefront of sustainable waste disposal is a commitment to waste reduction at its source. The crux of this approach lies in minimizing the generation of waste materials through conscious consumer choices, product design, and lifestyle practices. Individuals can significantly curb the influx of waste into the system by advocating for a reduction in single-use items, opting for durable and reusable products, and supporting businesses with eco-friendly packaging.

Source separation, or sorting waste at its origin, is pivotal in streamlining recycling. Households, businesses, and communities adopting efficient source separation systems facilitate recycling materials such as paper, glass, metal, and plastic. Segregating waste at the source not only simplifies the recycling process but also ensures that materials are channeled toward appropriate treatment methods, thereby enhancing the efficacy of waste disposal.

Recycling, a cornerstone of sustainable waste management, breathes new life into discarded materials by transforming them into raw materials to produce new goods. The recycling process minimizes the extraction of virgin resources, conserves energy, and diminishes the environmental impact associated with the manufacturing of products from scratch. Communities prioritizing recycling contribute to establishing a circular economy—an economic model emphasizing the continuous circulation of resources through recycling and reusing, as opposed to the traditional linear model of take, make, and dispose of.

Materials such as paper, glass, metal, and certain plastics undergo recycling processes that vary in complexity. While paper can be recycled into new paper products, glass, and metal can be melted down and reformed into containers or other items. Plastics, marked by their diverse compositions, pose unique challenges, yet advancements in recycling technologies strive to address these complexities and divert plastics from landfills and oceans.

Composting emerges as a sustainable solution for managing organic waste, encompassing kitchen scraps, yard waste, and other biodegradable materials. Through the natural decomposition of organic matter by microorganisms, composting generates nutrient-rich soil amendments that enhance soil fertility and structure. Adopting composting systems by individuals, communities, and even large-scale operations aligns with the principles of regenerative agriculture and contributes to the closing of organic nutrient cycles.

Beyond traditional composting, innovations in organic waste management include anaerobic digestion, which breaks down organic matter without oxygen, producing biogas and nutrient-rich slurry. Anaerobic digestion addresses the challenge of organic waste disposal and harnesses renewable energy in biogas, which can be utilized for heating, electricity generation, or as a clean cooking fuel.

Waste-to-energy technologies provide an avenue for the sustainable disposal of certain types of waste, particularly those that are not easily recyclable or compostable. Incineration, a form of waste-to-energy, involves the controlled burning of waste materials to generate heat, which is then used to produce electricity or heat buildings. While incineration can be a viable option for reducing the volume of waste and generating energy, concerns about air pollution, greenhouse gas emissions, and the potential release of harmful substances necessitate strict regulatory measures and advanced pollution control technologies.

Additionally, advancements in waste-to-energy technologies include gasification and pyrolysis, which involve the thermal breakdown of waste materials without oxygen. These processes yield syngas or bio-oil, which can be used as fuels, along with biochar—a stable carbon- rich material that enhances soil fertility and sequesters carbon.

Landfills, despite being a standard method of waste disposal, pose significant environmental challenges, including the release of greenhouse gases and the potential for groundwater contamination. Sustainable landfill management strategies involve the implementation of engineered landfills that incorporate liners, leachate collection systems, and gas collection systems to minimize environmental impact. Additionally, landfill mining—an innovative approach—aims to extract valuable materials from existing landfills, divert reusable resources, and mitigate the long-term ecological consequences of conventional landfill practices.

Landfill remediation involves the rehabilitation and restoration of decommissioned or poorly managed landfills. By implementing ecological restoration techniques, such as planting native vegetation and restoring natural habitats, these sites can be transformed into environmentally friendly spaces, contributing to biodiversity and ecosystem health.

While sustainable waste disposal methods offer promising solutions to the challenges posed by escalating waste generation, several considerations and challenges must be navigated. The efficacy of recycling programs depends on public participation, awareness, and adherence to source separation practices. The infrastructure for waste collection, recycling facilities, and waste-to-energy technologies must be developed and maintained to ensure the seamless integration of sustainable waste disposal methods.

Moreover, the global nature of waste trade poses ethical and environmental dilemmas, as developed nations export their waste to less economically developed regions. Addressing this issue requires international cooperation, policy interventions, and the development of circular economies that prioritize waste reduction and responsible waste management.

In conclusion, sustainable waste disposal methods are crucial components of a responsible and environmentally conscious approach to waste management. From waste reduction and recycling to composting, waste-to-energy technologies, and landfill management, a spectrum of options exists for individuals, communities, and policymakers to adopt. The collective transition towards sustainable waste disposal represents a paradigm shift that embraces resource efficiency, environmental harmony, and the preservation of ecosystems. As the global community grapples with the urgency of mitigating climate change and safeguarding natural resources, adopting sustainable waste disposal practices emerges as a pivotal step towards a more resilient and sustainable future.

Minimalist living practices

Amid a modern era marked by consumerism, material excess, and the constant pursuit of more, minimalist living practices have emerged as a counter-cultural movement, inviting individuals to embrace a lifestyle characterized by intentional simplicity and mindful consumption. Rooted in the philosophy that less is more, minimalist living transcends mere decluttering and organization; a profound shift in mindset prompts individuals to reevaluate their priorities, question societal norms, and foster a deeper connection to the essence of life. This section explores the fundamental principles of minimalist living, delving into its profound impact on individuals' well-being, the environment, and the broader social fabric.

At its core, minimalist living is a conscious decision to prioritize experiences, relationships, and personal growth over accumulating possessions. It challenges the prevailing notion that happiness and fulfillment are intrinsically tied to the possession of material goods. By intentionally simplifying one's life, individuals embark on a transformative journey that encourages them to question the societal narrative that equates success with accumulating wealth and belongings.

The decluttering aspect of minimalism involves assessing possessions with a discerning eye, retaining only those items that serve a purpose or bring genuine joy. This process, popularized by Marie Kondo's KonMari method, extends beyond a mere tidying-up exercise; it becomes a deliberate practice of curating one's environment to reflect personal values and priorities. As individuals let go of excess belongings, they often experience a profound sense of liberation, shedding the weight of unnecessary material burdens and creating physical and mental space for what truly matters.

Minimalism extends beyond the confines of one's living space to encompass various aspects of life, including relationships, work, and leisure. In relationships, minimalist living encourages individuals to nurture meaningful connections, focusing on quality over quantity. This intentional approach to social interactions promotes a more profound sense of community and shared experiences, fostering a supportive network that transcends the superficial.

In the realm of work, minimalist principles challenge the prevailing culture of overwork and excessive busyness. Rather than measuring success by the volume of tasks accomplished or possessions acquired, minimalist living encourages individuals to prioritize meaningful and fulfilling work that aligns with their values. This shift towards intentional and purpose-driven careers often leads to greater job satisfaction and a more balanced life.

The minimalist ethos also extends to leisure and recreation, prompting individuals to engage in activities that bring genuine joy and align with their passions. In a world inundated with distractions and constant stimuli, minimalist living advocates for a mindful approach to leisure, where quality prevails over quantity. This might involve savoring a quiet moment in nature, immersing oneself in a meaningful book, or pursuing hobbies that ignite genuine passion.

From an environmental perspective, minimalist living aligns with sustainability and conscious consumption principles. The fast-paced cycle of production and consumption that characterizes contemporary society contributes to resource depletion, environmental degradation, and a culture of waste. Minimalism challenges this culture by promoting the thoughtful consideration of the ecological impact of one's choices, emphasizing the importance of reducing, reusing, and recycling.

Decluttering and minimizing possessions inherently reduces the demand for new goods, contributing to a more sustainable and eco-friendly lifestyle. By investing in high-quality, durable items rather than succumbing to the allure of disposable and trend-driven consumer goods, minimalists actively reduce their ecological footprint. Moreover, the rejection of excessive packaging and single-use items and the constant pursuit of the latest trends align with the ethos of mindful consumption, fostering a sense of environmental responsibility.

The minimalist approach to living also has profound implications for personal finances. Individuals practicing minimalism often pursue financial freedom by prioritizing needs over wants and curbing impulsive spending. The intentional focus on what truly brings value and joy can lead to significant savings, allowing for the pursuit of meaningful experiences, investments in personal development, or the ability to contribute to social causes.

Minimalist living practices have challenges. In a culture that often equates success with material wealth and possessions, embracing a minimalist lifestyle can be met with skepticism or resistance. The pressure to conform to societal expectations and the fear of being judged for deviating from the norm can create internal conflicts for those embarking on the minimalist journey. Additionally, the emotional attachment to possessions, fueled by sentimental value or societal conditioning, can pose a significant hurdle in decluttering and simplifying.

However, it is precisely in navigating these challenges that the transformative power of minimalism becomes evident. The journey towards a minimalist lifestyle involves self-reflection, introspection, and reevaluating one's values. It invites individuals to confront societal narratives, question the pursuit of external validation, and rediscover a sense of authenticity and purpose.

Minimalist living also intersects with the growing movement towards conscious consumerism and ethical choices. The rejection of fast fashion, disposable goods, and products associated with exploitative labor practices aligns with the values of social responsibility and ethical consumption. Minimalists often support businesses and brands prioritizing sustainability, fair labor practices, and transparency in their operations.

The impact of minimalist living extends beyond the individual to influence broader societal trends. As more individuals embrace the minimalist ethos, a potential ripple effect challenges the prevailing consumerism culture. This shift in collective consciousness can influence businesses to reevaluate their practices, prompting a move towards sustainable production, ethical sourcing, and focusing on products that stand the test of time.

In conclusion, minimalist living practices represent a powerful antidote to the prevailing culture of excess and perpetual dissatisfaction. Individuals embark on a journey towards greater fulfillment, purpose, and environmental responsibility by intentionally simplifying one's life. Minimalism is not a rigid set of rules but a guiding philosophy that invites individuals to question the status quo, prioritize what truly matters, and live in alignment with their values. As the world grapples with environmental sustainability challenges, mental well-being, and societal discontent, the minimalist movement emerges as a beacon of hope—a pathway towards a more intentional, meaningful, and harmonious way of life.

Reducing environmental impact

In the face of escalating environmental challenges, the imperative to reduce our collective environmental impact has become more urgent than ever. The global community is grappling with the consequences of climate change, loss of biodiversity, and the depletion of natural resources. Amidst this reality, individuals, communities, and businesses increasingly recognize the need to adopt sustainable practices promoting environmental stewardship. This section explores the multifaceted approaches to reducing environmental impact, emphasizing the interconnectedness of individual choices, societal behavior, and systemic change in forging a more sustainable future.

At the heart of reducing environmental impact lies the recognition that human activities, from production and consumption to waste generation, have far-reaching consequences for the planet. Sustainable living entails a shift in mindset, a departure from the prevailing culture of convenience and disposability toward a more intentional and mindful approach to daily life. This shift encompasses a spectrum of practices, each contributing to the broader goal of minimizing environmental harm.

One of the critical facets of reducing environmental impact involves sustainable consumption. In a consumer-driven society, individuals wield significant influence through their purchasing choices. Opting for products with minimal packaging, prioritizing items with a longer lifespan, and favoring goods produced through ethical and environmentally responsible practices are pivotal steps in curbing the environmental repercussions of consumption. This approach aligns with the ethos of conscious consumerism, where individuals leverage their buying power to support businesses committed to sustainability and responsible sourcing.

Embracing energy efficiency is another crucial component of environmental impact reduction. Energy production and consumption significantly contribute to greenhouse gas emissions and environmental degradation. Implementing energy-efficient technologies like LED lighting, smart thermostats, and energy-efficient appliances can lower individual and collective carbon footprints substantially. Furthermore, the transition to renewable energy sources, such as solar and wind power, offers a transformative pathway toward a more sustainable and carbon-neutral energy landscape.

Waste reduction and responsible waste management are pivotal in minimizing environmental impact. The prevailing culture of single-use items and disposable packaging has led to alarming levels of plastic pollution, overflowing landfills, and adverse effects on ecosystems. Adopting a circular economy mindset, emphasizing reducing, reusing, and recycling, helps break free from the linear model of take, make, and dispose. Individuals can contribute by avoiding single-use plastics, repurposing items, and supporting community recycling initiatives.

Transportation choices significantly influence personal carbon footprints. The reliance on fossil fuel-powered vehicles contributes to air pollution, greenhouse gas emissions, and dependence on finite resources. Transitioning to eco-friendly modes of transportation, such as electric vehicles, bicycles, or public transit, is a tangible way to reduce the environmental impact of personal mobility. Additionally, carpooling and ride- sharing initiatives promote resource efficiency and contribute to the reduction of overall vehicle emissions.

Sustainable agriculture practices are essential in addressing the environmental impact of food production. Conventional agricultural methods, characterized by heavy pesticide use, monoculture, and excessive water consumption, contribute to soil degradation and biodiversity loss. Embracing organic farming, permaculture, and regenerative agriculture fosters a more harmonious relationship between food production and the environment. Local and seasonal food consumption further reduces the carbon footprint associated with transportation and storage.

The built environment significantly contributes to environmental impact, with energy-intensive construction materials and inefficient building designs exacerbating ecological strain. Embracing sustainable architecture and eco-friendly construction materials minimizes resource extraction, lowers energy consumption, and promotes energy-efficient structures. Passive solar design principles, green roofs, and the integration of renewable energy systems contribute to creating environmentally conscious buildings that align with the principles of reducing environmental impact.

Education and awareness are fundamental pillars in the quest to reduce environmental impact. Empowering individuals with knowledge about the ecological consequences of their choices fosters a sense of environmental responsibility. Educational initiatives on sustainable practices, biodiversity conservation, and climate change are crucial in inspiring informed decisions and cultivating a collective commitment to environmental stewardship.

At the societal level, advocacy and community engagement catalyze systemic change. Raising awareness about environmental issues, promoting sustainable policies, and supporting eco-friendly initiatives contribute to a broader cultural shift toward sustainability. Grassroots movements, environmental organizations, and community-based projects demonstrate the power of collective action in influencing policy changes and fostering a culture of environmental responsibility.

Corporate responsibility and sustainable business practices are instrumental in shaping the trajectory of environmental impact reduction. Companies embracing eco-friendly initiatives, adopting sustainable supply chain practices, and prioritizing environmental, social, and governance (ESG) criteria contribute to a more sustainable global economy. Sustainable business practices reduce negative ecological externalities and foster innovation, resilience, and long-term viability.

Despite progress in various arenas, challenges persist in reducing environmental impact. Economic incentives that prioritize short-term gains over long-term sustainability, inadequate regulatory frameworks, and the persistence of unsustainable practices in specific industries pose formidable obstacles. Overcoming these challenges requires a concerted effort from individuals, businesses, governments, and international organizations to prioritize the planet's health over immediate economic interests.

In conclusion, reducing environmental impact is an imperative that transcends individual actions; it requires a holistic and interconnected approach. From sustainable consumption and energy efficiency to waste reduction, responsible transportation, and eco-friendly construction, every facet of human activity shapes the environmental landscape. A collective commitment to education, advocacy, and systemic change is essential for creating a sustainable future where humanity coexists harmoniously with the planet. As the global community grapples with the urgency of addressing climate change and environmental degradation, embracing sustainable practices becomes a choice and a moral responsibility—a pledge to safeguard the Earth for current and future generations.

CHAPTER VII

Smart Technology Integration

Off-grid communication solutions

In the dynamic landscape of off-grid living, where individuals and communities embrace self-sufficiency and independence from traditional infrastructure, effective communication becomes both a challenge and a necessity. Off-grid environments, characterized by their remoteness and lack of conventional connectivity, demand innovative solutions to bridge communication gaps. This section explores the diverse array of off-grid communication solutions, from traditional methods adapted for modern needs to cutting-edge technologies redefining connectivity in remote settings.

Traditional methods serve as the foundation for connectivity in off-grid scenarios, where reliance on centralized power and communication grids is limited. Two-way radios, often powered by rechargeable batteries or solar panels, enable short-range communication between individuals within a defined area. These radios, with features like weather channels and emergency frequencies, are valuable for coordination and safety in off-grid communities.

Satellite communication represents another traditional yet reliable option for off-grid connectivity. Satellite phones, equipped with antennas for direct communication with orbiting satellites, provide voice and data communication in remote areas where conventional cellular networks are absent. While satellite communication may have higher upfront costs and limitations regarding data speed, it remains a lifeline for those in genuinely isolated regions.

The advent of technology has seen the evolution and integration of traditional tools into modern off-grid communication solutions. Two-way radio systems, enhanced by digital technology, offer improved clarity, range, and additional features such as text messaging and GPS tracking. These advancements make digital two- way radios indispensable for coordinating activities, managing security, and ensuring efficient communication in off-grid settings.

Mesh networks, a decentralized form of communication, leverage interconnected devices to create a web of communication nodes. Each device in the mesh network serves as a relay point, allowing messages to hop from one node to another until reaching the intended recipient. This autonomous and adaptable network is particularly effective in off-grid environments, where the lack of centralized infrastructure necessitates creative solutions for maintaining connectivity.

The off-grid lifestyle often aligns with sustainable practices; solar-powered communication solutions exemplify this synergy. Solar chargers and panels provide a renewable energy source for powering communication devices, reducing the reliance on traditional electricity grids. This ensures continuous communication in off-grid settings and aligns with the broader ethos of minimizing environmental impact.

Solar-powered Wi-Fi hotspots and routers enable off-grid communities to establish local networks, facilitating internet connectivity even in remote locations. These setups, often integrated with energy-efficient technologies, contribute to the sustainability of off-grid communication solutions. Combining solar power and communication technology empowers individuals to stay connected while minimizing their ecological footprint.

In recent years, advancements in satellite technology have given rise to satellite internet services that cater specifically to off-grid and remote environments. Low Earth Orbit (LEO) satellite constellations, comprising numerous small satellites orbiting closer to the Earth, have paved the way for high-speed and low-latency satellite internet. Companies deploying LEO satellites aim to provide global coverage, making internet access feasible even in the most isolated regions.

Satellite internet terminals with compact antennas enable off-grid users to access broadband internet services. These terminals, powered by solar energy or other renewable sources, connect to LEO satellites to establish a reliable internet connection. Such solutions bridge communication gaps and open avenues for online education, telemedicine, and access to information in off-grid communities.

Innovative solutions in off-grid communication extend to uncrewed aerial vehicles (UAVs) or drones. Meshed drone networks can serve as communication relays in areas with challenging terrain or limited infrastructure. Drones with communication devices can fly over obstacles, establishing temporary communication links between remote locations.

These drone networks are precious in emergencies, where traditional communication infrastructure may be disrupted. They offer a rapid response mechanism for establishing communication in disaster-stricken or isolated areas, providing a lifeline for coordination, rescue efforts, and information dissemination.

While off-grid communication solutions present a spectrum of possibilities, challenges persist in implementing and sustaining these technologies. Limited bandwidth, particularly in satellite internet services, may result in data usage restrictions and affect internet connectivity speed. Additionally, the upfront costs of specific technologies, such as satellite terminals, may pose financial barriers for individuals or communities in off-grid settings.

While sustainable, reliance on solar power introduces challenges related to weather conditions and seasonal variations in sunlight. Cloudy days or prolonged periods of low sunlight can impact the efficiency of solar-powered communication solutions. Addressing these challenges requires a combination of technological advancements, community engagement, and innovative financing models to make these solutions more accessible.

The successful implementation of off-grid communication solutions hinges on community engagement, capacity building, and the empowerment of individuals to use and maintain these technologies. Training programs that teach residents how to operate and troubleshoot communication devices, harness solar power, and establish local networks are critical components of sustainable off-grid communication strategies.

Moreover, community involvement in designing and deploying communication solutions ensures that these technologies align with the specific needs and contexts of off-grid living. Tailoring solutions to local requirements fosters a sense of ownership and encourages the long-term sustainability of communication initiatives in off-grid communities.

Off-grid communication solutions represent a dynamic intersection of tradition and innovation, addressing the unique challenges of remote and self-sufficient living. From traditional two-way radios and satellite communication to modern adaptations like mesh networks and solar-powered Wi-Fi, the spectrum of available technologies offers diverse options for staying connected in off-grid environments.

As technology evolves, off-grid communication solutions will likely become more sophisticated, affordable, and accessible. The integration of renewable energy sources advances in satellite technology, and the emergence of innovative concepts like meshed drones contribute to a future where geographical constraints do not bind connectivity.

In pursuing resilient and sustainable off-grid communication, it is essential to navigate challenges collaboratively, engage communities, and ensure that these technologies contribute to the well-being and empowerment of individuals in off-grid settings. By embracing the possibilities offered by off-grid communication solutions, we pave the way for a connected, informed, and resilient future in even the world's most remote corners.

High-tech solutions for sustainable living

In the ongoing pursuit of sustainable living, the integration of high-tech solutions has emerged as a transformative force, promising a harmonious coexistence between human activities and the planet. As the global community grapples with environmental challenges and the imperative to reduce ecological footprints, cutting-edge technologies offer a spectrum of possibilities to reshape how we live, consume resources, and interact with our surroundings. This section explores the role of high-tech solutions in fostering sustainable living, spanning diverse domains from energy and transportation to agriculture and urban planning.

One of the cornerstones of high-tech sustainable living is the implementation of smart grids. This sophisticated energy management system leverages digital technology to optimize electricity generation, distribution, and consumption. Intelligent grids integrate real-time data, advanced sensors, and communication networks to enhance the efficiency and reliability of energy delivery. By facilitating two-way communication between utilities and consumers, intelligent grids enable dynamic adjustments in response to demand fluctuations, contributing to a more resilient and sustainable energy infrastructure.

Energy efficiency, a pivotal aspect of sustainable living, is further augmented by smart home technologies. Smart thermostats, connected appliances, and home automation systems allow users to monitor and control energy usage in real time. Machine learning algorithms optimize heating, cooling, and lighting based on user preferences and behavioral patterns, reducing energy waste and promoting a more sustainable lifestyle.

High-tech solutions play a pivotal role in harnessing the vast potential of renewable energy sources. Advanced solar technologies, such as solar photovoltaic (PV) cells and concentrated solar power systems, continue to evolve, enhancing energy capture efficiency and lowering costs. Smart solar grids, capable of managing distributed solar generation and storage, enable seamless integration of solar energy into existing power grids.

Wind energy, another key player in the transition to renewable sources, benefits from technological advancements in turbine design and control systems. High-efficiency turbines with variable speed control and machine learning algorithms maximize energy extraction from the wind, contributing to the sustainability of wind power projects. Additionally, innovations in energy storage technologies, including advanced battery systems and grid-scale storage solutions, address the intermittent nature of renewable energy sources, ensuring a consistent and reliable power supply.

The electrification of transportation represents a paradigm shift towards sustainable mobility, driven by high-tech advancements in electric vehicles (EVs) and associated infrastructure. Electric cars, buses, and bikes, equipped with cutting-edge battery technologies and energy-efficient drivetrains, offer a cleaner and quieter alternative to traditional internal combustion engine vehicles. Integrating intelligent charging stations and grid-responsive charging management systems optimizes the use of renewable energy sources for powering EVs, reducing the environmental impact of transportation.

Autonomous and connected vehicles, enabled by artificial intelligence (AI) and sensor technologies, further enhance the efficiency and safety of transportation systems. These technologies promote streamlined traffic flow, reduce congestion, and minimize energy consumption through optimized route planning. The advent of shared mobility services, facilitated by ride-hailing platforms and electric, autonomous fleets, contributes to a more resource-efficient and sustainable urban transport ecosystem.

In agriculture, high-tech solutions are revolutionizing traditional farming practices, fostering sustainable approaches through precision agriculture. Remote sensing technologies, including satellite imagery and drones, provide farmers real-time data on crop health, soil moisture, and pest infestations. This granular information enables precise and targeted interventions, minimizing the need for chemical inputs and optimizing resource use.

IoT-enabled innovative farming systems integrate sensors, actuators, and data analytics to create a connected and responsive agricultural environment. Automated irrigation systems, precision planters, and AI-driven decision support systems enhance resource utilization efficiency, reduce water and chemical usage, and promote sustainable land management practices. These technologies contribute to increased agricultural productivity and minimize the environmental impact of conventional farming methods.

In the quest for sustainable living, smart cities have gained prominence, envisioning urban environments where high-tech solutions converge to enhance efficiency, resilience, and quality of life. IoT sensors, ubiquitous connectivity, and data analytics enable the creation of responsive and adaptive urban infrastructures. Smart grids, intelligent transportation systems, and energy-efficient buildings form an integrated framework that reduces resource consumption, lowers emissions, and enhances sustainability.

Urban planning informed by data-driven insights facilitates optimized land use, efficient public services, and the creation of green spaces. Advanced waste management systems, incorporating IoT sensors and automation, contribute to waste reduction, recycling, and circular economy practices. Innovative city initiatives also prioritize the well-being of residents by promoting pedestrian-friendly spaces, public transit, and intelligent healthcare solutions.

While high-tech solutions offer promising avenues for sustainable living, they are not without challenges and ethical considerations. The digital divide, characterized by disparities in access to technology and information, raises concerns about social equity and inclusion. Ensuring that the benefits of high-tech sustainability solutions reach all segments of society requires proactive measures to bridge the digital gap and address issues of affordability and accessibility.

Cybersecurity considerations are paramount in the deployment of interconnected technologies. The reliance on digital systems and data sharing poses potential risks, including privacy breaches and vulnerabilities to cyber threats. Implementing robust cybersecurity measures, ethical data practices, and transparent governance frameworks is essential for safeguarding the integrity and security of high-tech sustainability solutions.

The environmental impact of manufacturing and disposing of high-tech devices poses another set of challenges. The extraction of rare earth metals, energy-intensive production processes, and electronic waste contribute to environmental degradation. Circular economy principles, emphasizing product durability, repairability, and recycling, are crucial for mitigating the ecological footprint of high-tech solutions.

In conclusion, high-tech solutions stand at the forefront of the drive toward sustainable living, offering innovative tools to address environmental challenges and transform human interactions with the planet. From smart grids and renewable energy integration to electric mobility, precision agriculture, and smart cities, the convergence of technology and sustainability presents a visionary path toward a more resilient and harmonious future.

As these high-tech solutions evolve, it is imperative to navigate challenges collaboratively, ensuring that the benefits of technological advancements are equitably distributed. Ethical considerations, including privacy, cybersecurity, and environmental impact, must guide the deployment of high-tech sustainability solutions, fostering a responsible and inclusive approach to innovation. By embracing the possibilities offered by high-tech solutions, society has the potential to usher in an era where cutting-edge technology catalyzes sustainable living, shaping a future where humanity thrives in harmony with the planet.

Smart appliances for energy efficiency

In pursuing a more sustainable future, integrating smart appliances into homes has emerged as a transformative strategy to enhance energy efficiency and minimize environmental impact. Intelligent appliances with advanced sensors, connectivity features, and artificial intelligence algorithms redefine how households consume and manage energy. This section explores the multifaceted role of intelligent appliances in promoting energy efficiency, contributing to a greener lifestyle, and reshaping the dynamics of modern households.

One of the fundamental contributions of intelligent appliances to energy efficiency lies in their ability to monitor and optimize energy consumption. These appliances, from smart thermostats and lighting systems to intelligent refrigerators and washing machines, provide real-time data on energy usage patterns. Homeowners can access this information through dedicated apps, allowing them to make informed decisions about when and how to use appliances more efficiently.

Smart thermostats exemplify this capability by learning user preferences, adjusting heating or cooling systems based on occupancy patterns, and optimizing energy consumption. Similarly, intelligent lighting systems can be programmed to adjust brightness and color temperature according to natural light conditions, occupancy, or user preferences, minimizing unnecessary energy consumption. These appliances empower users with insights into their energy usage, fostering a culture of awareness and responsible consumption.

Smart appliances play a crucial role in demand response programs, which aim to balance energy demand with supply by adjusting consumption during peak periods or in response to grid conditions. Appliances such as intelligent water heaters, air conditioners, and electric vehicles can participate in demand response initiatives, temporarily reducing energy usage during high demand or grid instability.

Through connectivity with intelligent grids and utility systems, these appliances receive signals to adjust their operation, contributing to the overall stability and efficiency of the energy grid. This demand-side management benefits the grid and often comes with financial incentives for consumers, encouraging the widespread adoption of smart appliances as part of a collective effort to enhance energy resilience.

The rise of renewable energy sources, such as solar and wind power, has prompted a shift towards decentralized energy generation. Intelligent appliances are well-positioned to complement this shift by integrating seamlessly with renewable energy systems. For instance, smart home energy management systems can coordinate the operation of appliances with the availability of solar energy, optimizing consumption during peak solar generation hours.

Additionally, intelligent storage solutions, such as energy storage systems and smart batteries, enable homeowners to store excess energy generated from renewables for later use. These systems can automatically charge during periods of abundant renewable energy and discharge during periods of high demand or when renewable generation is low. The synergy between intelligent appliances and renewable energy sources contributes to a more sustainable and resilient energy ecosystem.

Incorporating machine learning and predictive analytics sets smart appliances apart by enabling them to adapt and optimize their performance based on user behavior and external factors. Advanced algorithms analyze historical usage patterns, weather forecasts, and user preferences to anticipate energy needs and maximize appliance operation for efficiency.

For example, an intelligent dishwasher with machine learning capabilities can adapt its cycle times based on historical usage data, load characteristics, and energy tariff information to run when electricity rates are lower or when renewable energy generation is high. This intelligent decision-making enhances energy efficiency and aligns with cost savings and environmental sustainability goals.

The concept of a smart home extends beyond individual appliances to encompass interconnected ecosystems where devices communicate with each other to optimize overall energy usage. A smart home hub or ecosystem acts as a central control point, allowing different appliances to share data and coordinate their operations for maximum efficiency.

For instance, a smart home ecosystem may enable coordinating heating, ventilation, and air conditioning (HVAC) systems with smart blinds and lighting to create a cohesive strategy for temperature regulation and natural light utilization. By leveraging data from multiple sources, these interconnected systems work synergistically to enhance energy efficiency and create a more comfortable living environment.

Smart appliances empower users with unprecedented control over their energy consumption through user-friendly interfaces and automation features. Mobile apps, voice-activated assistants, and centralized control panels allow homeowners to monitor and manage their appliances remotely, optimizing settings for energy efficiency.

Automation features enable the creation of customized schedules and scenarios, allowing users to program appliances according to their preferences and daily routines. For example, smart thermostats can automatically adjust temperatures based on occupancy patterns, ensuring energy is not wasted on unoccupied heating or cooling spaces. This level of user control fosters a sense of empowerment and engagement with energy-efficient practices.

While the potential benefits of intelligent appliances for energy efficiency are substantial, specific challenges and considerations must be addressed for widespread adoption. Interoperability and standardization of communication protocols remain critical to ensuring seamless connectivity among diverse smart devices and platforms. This interoperability is essential for the creation of cohesive smart home ecosystems where appliances can communicate effectively with one another.

Privacy and cybersecurity concerns loom as smart appliances gather and transmit sensitive data about user behaviors and preferences. Robust security measures, including encryption and secure authentication, are imperative to protect user privacy and prevent unauthorized access to smart home systems. Additionally, manufacturers must prioritize the longevity and upgradability of intelligent appliances to minimize electronic waste and ensure that devices remain relevant as technology evolves.

Smart appliances for energy efficiency represent a pivotal advancement in the quest for sustainable living, reshaping how households interact with energy resources. From energy monitoring and demand response to integration with renewable sources and the application of machine learning, these appliances offer a holistic approach to optimizing energy usage. As smart home ecosystems continue to evolve and user adoption increases, the collective impact of intelligent appliances on energy efficiency holds the promise of a more sustainable and resilient future. The ongoing innovation in this field contributes to individual comfort and convenience and aligns with the broader imperative of mitigating climate change and fostering a responsible and energy-conscious society.

Automation for off-grid homes

In the realm of off-grid living, where self-sufficiency and sustainability take center stage, the integration of automation technologies emerges as a game-changer. Automation, driven by advancements in intelligent systems, sensors, and artificial intelligence, offers off-grid homeowners the tools to enhance efficiency, manage resources, and elevate the overall quality of life. This section explores how automation is transforming off-grid homes, from energy management and resource optimization to facilitating a more comfortable and secure living environment.

At the heart of automation in off-grid homes lies the ability to intelligently manage and optimize energy usage. Off-grid living often relies on renewable energy sources, such as solar panels or wind turbines, coupled with energy storage solutions. Automation technologies allow for the seamless integration of these components, enabling intelligent energy management systems to monitor, control, and optimize the generation and consumption of energy.

For instance, intelligent inverters paired with solar panels can dynamically adjust the amount of energy fed into the off-grid power system based on real-time factors such as weather conditions and energy demand. Battery storage systems, enhanced by automation, can optimize charging and discharging cycles, ensuring that stored energy is efficiently utilized. These technologies contribute to a more reliable and sustainable off-grid energy infrastructure, reducing reliance on conventional power grids.

Automation extends its reach to efficiently utilizing resources beyond energy, encompassing water, heating, and cooling systems in off-grid homes. Intelligent water management systems with sensors and actuators can monitor water usage patterns, detect leaks, and automate irrigation processes in gardens or agricultural areas. This conserves water and ensures that this precious resource is used judiciously in off-grid environments where water sources may be limited.

Heating, ventilation, and air conditioning (HVAC) systems in off-grid homes can benefit from automation for temperature regulation based on occupancy patterns and external weather conditions. Smart thermostats with learning algorithms adapt to user preferences and adjust heating or cooling systems for maximum efficiency. These technologies contribute to a comfortable living environment while minimizing energy waste.

Automation enhances the security and surveillance aspects of off-grid living, providing homeowners with advanced tools to monitor and protect their properties. Intelligent security systems, incorporating sensors, cameras, and motion detectors, can be integrated into a centralized automation platform. This allows homeowners to monitor their premises remotely, receive real-time alerts, and even automate responses to potential security threats.

Automation in off-grid security systems may include features such as smart locks, lighting control, and automated gates. For instance, motion sensors can trigger outdoor lights to illuminate dark areas, serving as a deterrent and enhancing visibility. Integrating these technologies not only enhances the security of off-grid homes but also contributes to residents' peace of mind, particularly in remote settings where traditional security infrastructure may be lacking.

Automating appliances and home control systems represents a cornerstone of intelligent living in off-grid environments. Smart home hubs or platforms serve as centralized control centers, allowing homeowners to seamlessly manage various devices and systems. From lighting and entertainment systems to kitchen appliances, automation enables users to control and monitor their homes conveniently and efficiently.

For example, a smart home system can automate the operation of lights and appliances based on occupancy or time of day, reducing unnecessary energy consumption. Remote control capabilities allow users to monitor and adjust settings even when away from their off-grid homes. This level of automation enhances energy efficiency and contributes to a more streamlined and interconnected living experience.

Automation technologies provide off-grid homeowners valuable insights into their living environment through data monitoring and analysis. Sensors deployed throughout the home can gather temperature, humidity, air quality, and energy usage data. This information is then processed and presented to users through user- friendly interfaces, offering a comprehensive understanding of the off-grid home's performance.

Data analysis enables homeowners to make informed decisions about resource management, energy efficiency improvements, and system optimizations. For instance, historical energy usage data can inform decisions about upgrading renewable energy systems or expanding energy storage capacity. Automation transforms data into actionable intelligence, continuously empowering residents to enhance their off-grid homes' sustainability and performance.

While integrating automation in off-grid homes brings many benefits, specific challenges and considerations must be addressed. Dependence on technology and connectivity raises concerns about system reliability, particularly in remote locations with limited access to maintenance and technical support. Off-grid homeowners must carefully select robust and durable automation systems to withstand the challenges of the off-grid lifestyle.

Power consumption by automation systems is another consideration, especially in environments where energy resources are scarce. Striking a balance between the benefits of automation and the energy required to operate smart devices is essential. Energy-efficient automation technologies, coupled with reasonable use and optimization, can mitigate this challenge and ensure that the benefits outweigh the costs.

In conclusion, automation is reshaping the landscape of off-grid living, offering tools to enhance efficiency, conserve resources, and elevate the overall living experience. From intelligent energy management and resource optimization to security, surveillance, and data-driven insights, automation technologies contribute to the self-sufficiency and sustainability of off-grid homes.

As technology advances, the integration of automation in off-grid living will likely become more sophisticated, user-friendly, and tailored to the unique challenges and opportunities presented by remote and self-sufficient environments. By harnessing the power of automation, off-grid homeowners can embrace a more sustainable lifestyle and enjoy the comforts and conveniences afforded by cutting-edge technology, fostering a harmonious relationship between human habitation and the natural world.

CHAPTER VIII

Community Building

Connecting with like-minded individuals

In the quest for sustainable and off-grid living, the significance of connecting with like-minded individuals cannot be overstated. Beyond the practical aspects of self-sufficiency, energy independence, and eco-friendly practices, the communal element is pivotal in shaping the off-grid lifestyle. This section delves into the profound impact of forming connections with like-minded individuals, exploring the benefits of shared values and mutual support, and creating resilient communities dedicated to a sustainable way of life.

The foundation of connecting with like-minded individuals lies in the shared values and common goals that underpin the off-grid lifestyle. Whether motivated by a commitment to environmental conservation, a desire for greater self-sufficiency, or a shared vision of reducing one's ecological footprint, like-minded individuals find common ground in their dedication to sustainable living. This shared ethos forms the basis for a community where members can draw inspiration, share experiences, and collectively contribute to realizing common objectives.

The importance of shared values extends beyond the practical aspects of off-grid living. It encompasses a deeper alignment of beliefs and principles that foster a sense of belonging and purpose. When individuals connect with like-minded peers who share their passion for sustainability, a supportive community emerges, creating a robust network of individuals united by a collective vision of a more harmonious relationship with the environment.

One of the inherent advantages of connecting with like-minded individuals in the off-grid community is the opportunity for mutual support and knowledge exchange. The journey toward sustainable living often involves navigating challenges related to renewable energy systems, permaculture practices, or off-grid construction. In a community of like-minded individuals, members can share their experiences, insights, and lessons learned, providing valuable guidance and support to one another.

This knowledge exchange extends beyond technical expertise, encompassing practical wisdom from real-world experiences. Whether it's troubleshooting a solar power system, optimizing water harvesting techniques, or sharing tips for sustainable gardening, the collective intelligence of a like-minded community becomes a valuable resource for individuals navigating the complexities of off-grid living. This mutual support enhances the resilience of individual households and contributes to the overall strength of the off-grid community.

Forming like-minded communities in the realm of sustainable and off-grid living goes beyond individual benefits; it contributes to creating resilient and interdependent social structures. Resilient communities are characterized by their ability to adapt to challenges, share resources, and collectively address common concerns. In the context of off-grid living, these communities become microcosms of sustainability, embodying the principles of self-sufficiency, cooperation, and environmental stewardship.

Resilient communities thrive on collaboration, where individuals with diverse skills and expertise come together to build a collective foundation of knowledge and capabilities. From shared food production to cooperative energy initiatives, these communities demonstrate the power of synergy in addressing the multifaceted aspects of off-grid living. The bonds formed within these communities foster a sense of security and belonging, laying the groundwork for a sustainable and fulfilling lifestyle.

Connecting with like-minded individuals extends beyond personal relationships to community-based initiatives that amplify the impact of shared values. Off-grid communities often collaborate on projects that benefit the collective, such as establishing community gardens, implementing renewable energy systems, or organizing workshops on sustainable practices. These initiatives contribute to the well-being of community members and serve as examples of sustainable living that can inspire broader societal change.

Community-based initiatives foster a sense of responsibility and shared stewardship of the environment. When individuals unite to address local challenges related to waste management, water conservation, or habitat restoration, they become agents of positive change within their communities. The ripple effect of these initiatives extends beyond the immediate neighborhood, influencing neighboring areas and contributing to a broader cultural shift toward sustainability.

The off-grid lifestyle, while rewarding, can sometimes lead to feelings of isolation, especially in remote settings. Connecting with like-minded individuals minimizes this isolation by creating a sense of community and shared purpose. Shared experiences, celebrations, and communal activities enhance well-being and counteract the potential challenges of solitude often associated with off-grid living.

The emotional support derived from a community of like-minded individuals becomes particularly crucial during challenging times, whether due to environmental factors, technical issues, or personal circumstances. The understanding and empathy within the community create a safety net that bolsters the emotional resilience of individuals, contributing to a more holistic and fulfilling off-grid lifestyle.

While connecting with like-minded individuals brings numerous benefits, specific challenges and considerations must be acknowledged. Geographical distance, particularly in remote off-grid settings, can pose logistical challenges for forming and maintaining communities. Overcoming these challenges may require creative solutions, such as online platforms, periodic gatherings, or collaborative initiatives that transcend physical proximity.

Diversity of perspectives within a like-minded community is another consideration. While shared values form the foundation, embracing diversity of thought and experience enriches the community by bringing different perspectives. Striking a balance between unity of purpose and openness to diverse viewpoints ensures a vibrant and dynamic community that can adapt to changing circumstances.

In conclusion, connecting with like-minded individuals is a cornerstone of the off-grid lifestyle, transforming the pursuit of sustainability into a communal endeavor. Shared values, mutual support, and the formation of resilient communities amplify the impact of individual efforts, contributing to a more sustainable and interconnected way of life. As individuals forge connections with like-minded peers, the off-grid community becomes a source of inspiration, knowledge, and collective strength, laying the groundwork for a future where sustainable living is not just a personal choice but a shared ethos that transcends individual households and resonates within resilient and thriving communities.

Sharing resources within a community

In pursuing sustainable and communal living, sharing resources within a community emerges as a cornerstone principle that transcends individual households and fosters a collective approach to environmental stewardship. This section explores the profound impact of resource sharing within communities, delving into the economic, ecological, and social dimensions that make collaborative sustainability a transformative force in sustainable living.

One of the primary benefits of sharing resources within a community lies in the economic efficiency and cost reduction that result from collective endeavors. Whether it's pooling financial resources for the purchase of shared equipment or sharing tools, machinery, or vehicles, communities that collaborate on resource utilization can achieve significant cost savings. This economic efficiency is particularly relevant in sustainable living, where the upfront costs of eco-friendly technologies and infrastructure can sometimes be a barrier for individual households.

Shared resource initiatives, such as community gardens or cooperative renewable energy projects, distribute the financial burden among community members, making sustainable practices more accessible. The shared responsibility for costs reduces the financial strain on individual households and promotes a sense of equity and inclusivity within the community. This way, collaborative sustainability catalyzes economic resilience and affordability in pursuing environmentally conscious living.

Sharing resources within a community extends beyond financial considerations to optimizing energy and water usage. In off-grid or sustainable communities, where energy independence is a common goal, collective efforts to harness renewable energy sources often lead to more efficient and cost-effective solutions. Shared solar arrays, wind turbines, or micro-hydro systems can be collectively owned and maintained, ensuring that the benefits of clean energy are distributed equitably among community members.

Water conservation is another area where resource sharing within a community can yield substantial benefits. Cooperative water harvesting and storage systems and shared irrigation practices in community gardens contribute to the sustainable use of water resources. By optimizing the collective management of energy and water, communities can reduce their environmental impact while enhancing the overall efficiency of resource utilization.

Sharing resources within a community aligns with the principles of waste reduction and the creation of circular economies. Communities that emphasize the reuse, repurposing, and sharing of goods contribute to a reduction in overall waste generation. Shared tool libraries, clothing swaps, and community repair workshops are examples of initiatives that promote the longevity of products and discourage the disposable culture prevalent in modern societies.

The concept of circular economies within communities revolves around closing the loop on resource consumption. Items that have reached the end of their lifecycle for one individual can find new purpose within the community through sharing and repurposing. This minimizes the environmental impact of resource extraction and manufacturing and cultivates a culture of mindful consumption and responsible waste management within the community.

One of the most tangible expressions of resource sharing within communities is seen in collaborative food production and agriculture initiatives. Community gardens, communal orchards, and shared agricultural spaces enable residents to pool their efforts and resources to cultivate fresh, locally-grown produce. Beyond the ecological benefits of reducing the carbon footprint associated with food transportation, these initiatives foster a shared responsibility for food security and self-sufficiency.

In addition to traditional farming practices, innovative approaches such as community-supported agriculture (CSA) models exemplify collaborative food production. CSAs involve community members directly in the agricultural process by subscribing to shares of a local farm's produce. This model provides residents with fresh and seasonal produce and establishes a direct link between consumers and producers, reinforcing the bonds of community and shared sustenance.

Resource sharing within a community goes beyond tangible benefits; it fosters social cohesion and interconnectedness among residents. Collaborative initiatives create opportunities for community members to engage with one another, share skills, and build relationships. Whether participating in a communal workday at a shared garden or collectively addressing challenges in resource management, residents become active contributors to the community's well-being.

This interconnectedness contributes to a sense of belonging and shared purpose, vital to a thriving community. Shared resources become a vehicle for strengthening social bonds, creating a supportive network that transcends individual households. In times of need or adversity, this sense of community becomes a source of resilience and mutual support, enriching the quality of life for all members.

While the benefits of sharing resources within a community are evident, specific challenges and considerations must be addressed to ensure the success and sustainability of collaborative initiatives. Communication and coordination are vital factors, as effective resource sharing requires clear communication channels and mechanisms for decision-making within the community. Establishing transparent resource allocation, maintenance, and ownership processes helps prevent potential conflicts and ensures equitable participation.

Cultural considerations and individual preferences also play a role in the success of resource-sharing initiatives. Recognizing and respecting diverse perspectives within the community is essential for fostering an inclusive environment. Flexibility in adapting shared resource models to accommodate varying needs and preferences contributes to the overall success and acceptance of collaborative sustainability practices.

In conclusion, sharing resources within a community is a transformative practice that enhances economic efficiency and fosters ecological sustainability and social interconnectedness. From optimizing energy and water usage to reducing waste and promoting collaborative agriculture, communities that embrace shared resource models embody the principles of sustainable living. As the global community grapples with environmental challenges, these local initiatives serve as inspiring examples of how collective action can contribute to a more resilient, equitable, and sustainable future. By weaving the fabric of collaborative sustainability into the daily lives of community members, shared resource practices become a powerful force for positive change, shaping a future where communities thrive in harmony with the environment.

Collaborative projects for sustainability

In the ongoing quest for a more sustainable and eco-conscious existence, the power of collaborative projects within communities has emerged as a dynamic force driving positive change. This section explores the multifaceted nature of joint projects for sustainability, examining their impact on environmental conservation, community resilience, and the collective pursuit of a greener future.

One of the most impactful collaborative projects for sustainability revolves around community-led renewable energy initiatives. In these projects, community members come together to harness clean and renewable energy sources, such as solar, wind, or hydropower, to reduce reliance on traditional energy grids. This collaborative approach promotes environmental stewardship and empowers communities to take control of their energy needs.

Community solar projects, for instance, involve installing solar panels in a shared location, allowing multiple households or businesses to benefit from clean energy generation. These initiatives often leverage economies of scale, making renewable energy technologies more accessible and affordable for participants. By collectively investing in and managing renewable energy infrastructure, communities can significantly reduce their carbon footprint and contribute to a more sustainable energy landscape.

Collaborative projects for sustainability extend to agriculture, where communities embrace cooperative and sustainable farming practices. Community-supported agriculture (CSA) models exemplify this approach, where individuals within a community collectively support local farmers by subscribing to seasonal shares of their produce. This collaborative model provides residents with fresh and locally sourced food and fosters a direct connection between consumers and producers.

Beyond CSAs, communal gardens and shared agricultural spaces allow community members to pool resources, expertise, and labor for the cultivation of crops. This collaborative approach to sustainable agriculture promotes food security, encourages biodiversity, reduces the ecological impact of large-scale farming practices, and strengthens the resilience of local food systems. In these projects, the soil becomes a canvas for collective action, fostering a sense of shared responsibility for the land and its bounty.

Collaborative projects aimed at waste reduction and recycling play a crucial role in fostering sustainable community practices. Recycling programs involving collective sorting and processing recyclable materials encourage responsible waste management. Community-wide initiatives, such as composting projects or material exchange programs, contribute to the circular economy by repurposing and reusing materials within the local ecosystem.

These projects often require active participation from community members, who may collectively organize clean-up events, recycling drives, or educational programs to raise awareness about waste reduction. By working together to minimize the environmental impact of waste, communities contribute to a cleaner and healthier local environment and set the stage for a cultural shift towards mindful consumption and waste reduction.

In urban settings, collaborative projects for sustainability often involve green infrastructure and community-driven urban planning initiatives. Residents, local governments, and environmental organizations may join forces to advocate for and implement green infrastructure projects such as urban parks, green roofs, and community gardens. These projects enhance the aesthetic appeal of urban areas and contribute to air quality improvement, stormwater management, and overall urban resilience.

Collaborative urban planning focuses on creating sustainable and walkable communities, promoting public transportation, and incorporating green spaces into urban development. By collectively shaping the urban environment, communities can reduce their ecological footprint, mitigate the heat island effect, and foster a sense of well-being among residents. These collaborative endeavors showcase the transformative potential of community-led initiatives in creating more sustainable and livable urban environments.

An integral aspect of collaborative sustainability projects involves education and awareness campaigns that seek to inform and engage community members. Whether through workshops, seminars, or community outreach programs, these initiatives empower individuals with knowledge about sustainable practices, environmental conservation, and the broader implications of collective action. Education catalyzes behavioral change, encouraging community members to adopt more sustainable lifestyles.

Collaborative education projects may include initiatives such as energy efficiency workshops, permaculture training, or climate change awareness campaigns. By fostering a culture of environmental literacy within the community, these projects amplify the impact of other sustainability initiatives, as informed individuals become catalysts for positive change within their households and broader social circles.

While collaborative projects for sustainability bring about numerous benefits, they also face challenges that require thoughtful consideration. One such challenge is effective communication and coordination among community members. Collaborative projects often involve diverse stakeholders with varying levels of engagement, priorities, and perspectives. Establishing transparent communication channels and inclusive decision-making processes is crucial to ensure the success and longevity of these initiatives.

Financial considerations may also pose challenges, especially in communities with diverse economic backgrounds. Ensuring equitable participation and addressing potential disparities in financial contributions requires careful planning and, in some cases, external support or grant funding. Overcoming financial barriers is essential to making collaborative sustainability projects accessible and inclusive for all community members.

In conclusion, collaborative projects for sustainability represent a powerful mechanism for communities to address environmental challenges, foster resilience, and collectively pursue a greener future. From renewable energy initiatives and sustainable agriculture to waste reduction programs and community-driven urban planning, these projects showcase the transformative potential of collective action. By harnessing the strength of community collaboration, individuals within these projects become architects of positive change, contributing to a more sustainable and resilient world where communities thrive in harmony with the environment. As communities continue to innovate and collaborate, the ripple effect of these collective efforts holds the promise of a future where sustainable practices are not only embraced but woven into the fabric of everyday life.

Building a resilient off-grid community

In sustainable living, off-grid communities have gained traction as individuals seek alternatives to conventional lifestyles. These communities, often nestled in remote or rural areas, prioritize self-sufficiency, environmental stewardship, and a harmonious relationship with nature. This section delves into the multifaceted aspects of building a resilient off-grid community, exploring the fundamental principles, challenges, and strategies that underpin the journey toward sustainable independence and collective strength.

The foundations of a resilient off-grid community are built on principles that prioritize sustainability, resourcefulness, and adaptability. At the core of this resilience is a commitment to reducing dependence on external resources, particularly conventional energy grids, water supplies, and food distribution systems. Off- grid communities often embrace renewable energy sources, such as solar, wind, or micro-hydro power, to generate electricity independently. Water self-sufficiency is achieved through rainwater harvesting, wells, or sustainable aquifer management, while food production is localized through community gardens, permaculture, and regenerative agriculture practices.

These foundational elements not only reduce the community's ecological footprint but also enhance its ability to withstand external shocks, such as disruptions in supply chains or fluctuations in energy prices. Resilience in the context of off-grid living involves cultivating a deep understanding of the local environment, harnessing natural resources judiciously, and fostering cooperation among community members to address challenges collectively.

Building resilience in an off-grid community necessitates a collaborative approach to decision-making. From the planning and establishment phases to daily operations and future developments, involving community members in decision-making fosters a sense of ownership and shared responsibility. Collaborative decision-making ensures that diverse perspectives are considered, allowing the community to leverage the collective wisdom of its members.

Decisions on energy infrastructure, water management, land use, and communal facilities are made through consensus-building processes, community meetings, or participatory forums. This inclusive approach strengthens the bonds within the community and promotes a shared vision and a sense of accountability for the community's well-being. Collaborative decision-making becomes a cornerstone of resilience by cultivating a culture of cooperation and adaptability.

Energy independence is a pivotal aspect of resilience in off-grid communities, and sustainable technologies play a central role in achieving this goal. Solar power systems, wind turbines, and micro-hydro systems harness renewable energy sources to generate electricity, providing off-grid communities with a reliable and environmentally friendly power supply. Battery storage systems enhance energy resilience by storing excess energy for use during low renewable energy production periods.

Integrating energy-efficient technologies, such as LED lighting, energy-efficient appliances, and intelligent energy management systems, further optimizes energy usage within the community. Off-grid communities embrace a holistic approach to energy independence, combining renewable energy sources with energy-efficient practices to minimize reliance on external grids and reduce the environmental impact of energy consumption.

Water scarcity is a common concern in off-grid settings, and building resilience involves implementing strategies for water self-sufficiency and conservation. Rainwater harvesting systems collect and store rainwater for various uses, including irrigation, gardening, and household needs. Well management and sustainable aquifer practices ensure a continuous and reliable water supply.

Water conservation practices, such as efficient irrigation methods, drought-tolerant plants, and community awareness campaigns, contribute to the sustainable management of water resources. Off-grid communities prioritize protecting and preserving local water sources, recognizing the integral role water plays in sustaining life and fostering resilience.

Resilient off-grid communities prioritize local food production through sustainable agricultural practices and permaculture principles. Community gardens, orchards, and permaculture designs maximize the use of available land, promote biodiversity, and minimize the environmental impact of food production. Permaculture principles, which emphasize design strategies that mimic natural ecosystems, guide the cultivation of resilient and regenerative agricultural systems.

Integrating agroforestry, rainwater harvesting for irrigation, composting, and organic fertilizers contributes to soil health and fertility. By cultivating diverse crops and integrating livestock for soil enrichment, off-grid communities enhance food security, reduce dependence on external food sources, and promote ecological resilience.

Resilient off-grid communities actively engage in waste reduction and circular economy practices to minimize their environmental footprint. Recycling initiatives, composting programs, and material repurposing projects contribute to the community's circular economy. By reducing, reusing, and recycling materials, these communities aim to minimize waste generation and promote a sustainable approach to resource management.

Cultivating a mindful consumption and waste reduction culture is often complemented by educational programs and community outreach efforts. Off-grid communities prioritize environmental awareness, encouraging residents to make informed choices that align with sustainability principles and circular economies.

Social resilience is a critical dimension of building a resilient off-grid community. Community-based initiatives, such as shared workdays, skill-sharing programs, and mutual support networks, contribute to the community's social fabric. These initiatives strengthen social bonds, foster a sense of belonging, and create a supportive environment where residents can rely on each other during times of need.

Participatory governance structures, community events, and shared celebrations contribute to the social resilience of off-grid communities. In times of challenges, such as extreme weather events or disruptions in essential services, the strong social networks within the community become a source of emotional support, practical assistance, and collective problem-solving.

Building a resilient off-grid community has, and careful consideration must be given to various factors. Access to resources, both financial and natural, can be a limiting factor for some communities. Overcoming economic barriers may require creative solutions, grant funding, or collaborative financing models that ensure equitable participation.

Geographical and climatic considerations also play a role in determining the viability and resilience of off-grid communities. Harsh climates, extreme weather events, or geological challenges may pose unique hurdles that require innovative solutions and adaptive strategies. Planning and designing infrastructure and systems that account for these challenges are essential for long-term resilience.

In conclusion, building a resilient off-grid community is a dynamic and multifaceted endeavor that addresses environmental, social, and economic dimensions. From energy independence and sustainable technologies to collaborative decision-making and community-based initiatives, resilience is cultivated through a holistic and inclusive approach. Off-grid communities, by embracing principles of sustainability and self-sufficiency, demonstrate the transformative potential of collective action in creating resilient and harmonious living environments. As the global community grapples with environmental challenges, the lessons learned from resilient off-grid communities inspire a future where sustainability, independence, and collective strength merge to shape a more resilient and sustainable world.

CONCLUSION

"Sustainable Visions: "Sustainable Visions: Off-Grid Projects for the Modern Homesteader - Elevate Your Lifestyle with Green Innovations" stands as a comprehensive and empowering guide, weaving a tapestry of knowledge for individuals seeking to embrace a sustainable and off-grid lifestyle. Through its exploration of diverse topics, from renewable energy systems and water harvesting to sustainable agriculture and community-building initiatives, the e-book provides practical insights. It fosters a vision for a more harmonious relationship with the environment.

The e-book is a beacon for those looking to elevate their

lifestyles through green innovations, offering a roadmap to independence, resilience, and eco-conscious living. Its emphasis on off-grid projects reflects a profound understanding of the interconnectedness of sustainable practices and the transformative impact they can have on individual households and entire communities.

The journey begins with a thoughtful assessment of land

for off-grid living, guiding readers through the crucial considerations that lay the groundwork for a self-sufficient and sustainable habitat. Legal concerns and permits are demystified, ensuring readers confidently navigate the regulatory landscape as they embark on their off-grid endeavors.

The e-book then delves into the heart of sustainable

living, exploring renewable energy systems, water harvesting techniques, and innovative approaches to shelter and construction. Each chapter is a gateway to a deeper understanding of the intricate web of choices available to modern homesteaders, providing the knowledge needed to make informed decisions that align with environmental stewardship.

As readers progress through the pages, the e-book expands beyond individual homesteads to explore the importance of community, collaboration, and shared values. It recognizes the power of like-minded individuals coming together to form resilient off-grid communities, amplifying the impact of sustainability efforts and creating a support network for those on a similar journey.

In its conclusion, "Sustainable Visions" not only

summarizes the wealth of information provided but also inspires a sense of possibility. It paints a vision of a future where individuals, equipped with the insights gained from the e-book, can thrive in self-sustaining and interconnected communities. Ultimately, the e-book invites readers to embark on a transformative journey, where the pursuit of green innovations becomes a catalyst for a lifestyle that is not just sustainable but one that elevates the very essence of modern homesteading.

Thank you for buying and reading/ listening to our book. If you found this book useful/ helpful please take a few minutes and leave a review on the platform where you purchased our book. Your feedback matters greatly to us.